WITHDRAWN FROM
UNIVERSITY OF BRIGHTON
LIBRARIES

KU-180-468

BN 9517021 9

Financial calculus

An introduction to derivative pricing

Financial calculus
An introduction to derivative pricing

Martin Baxter
University of Cambridge

Andrew Rennie
Union Bank of Switzerland

PUBLISHED BY THE PRESS SYNDICATE OF THE UNIVERSITY OF CAMBRIDGE
The Pitt Building, Trumpington Street, Cambridge CB2 1RP, United Kingdom

CAMBRIDGE UNIVERSITY PRESS
The Edinburgh Building, Cambridge CB2 2RU, United Kingdom
40 West 20th Street, New York, NY 10011-4211, USA
10 Stamford Road, Oakleigh, Melbourne 3166, Australia

© M.W. Baxter and A.J.O. Rennie 1996

This book is in copyright. Subject to statutory exception
and to the provisions of relevant collective licensing agreements,
no reproduction of any part may take place without
the written permission of Cambridge University Press

First published 1996

Reprinted with corrections 1997

Printed in the United Kingdom at the University Press, Cambridge

Typeset in Monotype Bembo by the authors using TEX

A catalogue record of this book is available from the British Library

ISBN 0 521 55289 3 hardback

9 517021

Contents

Contents

Preface

Notoriously, works of mathematical finance can be precise, and they can be comprehensible. Sadly, as Dr Johnson might have put it, the ones which are precise are not necessarily comprehensible, and those comprehensible are not necessarily precise.

But both are needed. The mathematics of finance is not easy, and much market practice is based on a soft understanding of what is actually going on. This is usually enough for experienced practitioners to price existing contracts, but often insufficient for innovative new products. Novices, managers and regulators can be left to stumble around in literature which is ill suited to their need for a clear explanation of the basic principles. Such 'seat of the pants' practices are more suited to the pioneering days of an industry, rather than the mature $15 trillion market which the derivatives business has become.

On the academic side, effort is too often expended on finding precise answers to the wrong questions. When working in isolation from the market, the temptation is to find analytic answers for their own sake with no reference to the concerns of practitioners. In particular, the importance of hedging both as a justification for the price and as an important end in itself is often underplayed. Scholars need to be aware of such financial issues, if only because some of the very best work has arisen in answering the questions of industry rather than academe.

Guide to the chapters

Chapter one is a brief warning, especially to beginners, that the expected worth of something is not a good guide to its price. That idea has to be shaken off and arbitrage pricing take its place.

Chapter two develops the idea of hedging and pricing by arbitrage in the discrete-time setting of binary trees. The key probabilistic concepts of conditional expectation, martingales, change of measure, and representation are all introduced in this simple framework, accompanied by illustrative examples.

Chapter three repeats all the work of its predecessor in the continuous-time setting. Brownian motion is brought out, as well as the Itô calculus needed to manipulate it, culminating in a derivation of the Black–Scholes formula.

Chapter four runs through a variety of actual financial instruments, such as dividend paying equities, currencies and coupon paying bonds, and adapts the Black–Scholes approach to each in turn. A general pattern of the distinction between tradable and non-tradable quantities leads to the definition the market price of risk, as well as a warning not to take that name too seriously. A section on quanto products provides a showcase of examples.

Chapter five is about the interest rate market. In spirit, a market of bonds is much like a market of stocks, but the richness of this market makes it more than just a special case of Black–Scholes. Market models are discussed with a joint short-rate/HJM approach, which lies within the general continuous framework set up in chapter three. One section details a few of the many possible interest rate contracts, including swaps, caps/floors and swaptions. This is a substantial chapter reflecting the depth of financial and technical knowledge that has to be introduced in an understandable way. The aim is to tell one basic story of the market, which all approaches can slot into.

Chapter six concludes with some technical results about larger and more general models, including multiple stock n-factor models, stochastic numeraires, and foreign exchange interest-rate models. The running link between the existence of equivalent martingale measures and the ability to price and hedge is finally formalised.

A short bibliography, complete answers to the (small) number of exercises, a full glossary of technical terms and an index are in the appendices.

How to read this book

The book can be read either sequentially as an unfolding story, or by random access to the self-contained sections. The occasional questions are to allow practice of the requisite skills, and are never essential to the development of the material.

A reader is not expected to have any particular prior body of knowledge, except for some (classical) differential calculus and experience with symbolic notation. Some basic probability definitions are contained in the glossary, whereas more advanced readers will find technical asides in the text from time to time.

Acknowledgements

We would like to thank David Tranah at CUP for politely never mentioning the number of deadlines we missed, as well as his much more invaluable positive assistance; the many readers in London, New York and various universities who have been subjected to writing far worse than anything remaining in the finished edition. Special thanks to Lorne Whiteway for his help and encouragement.

Martin Baxter   *June 1996*
Andrew Rennie

The parable of the bookmaker

A bookmaker is taking bets on a two-horse race. Choosing to be scientific, he studies the form of both horses over various distances and goings as well as considering such factors as training, diet and choice of jockey. Eventually he correctly calculates that one horse has a 25% chance of winning, and the other a 75% chance. Accordingly the odds are set at 3–1 against and 3–1 on respectively.

But there is a degree of popular sentiment reflected in the bets made, adding up to $5 000 for the first and $10 000 for the second. Were the second horse to win, the bookmaker would make a net profit of $1667, but if the first wins he suffers a loss of $5000. The expected value of his profit is $25\% \times (-\$5000) + 75\% \times (\$1667) = \$0$, or exactly even. In the long term, over a number of similar but independent races, the law of averages would allow the bookmaker to break even. Until the long term comes, there is a chance of making a large loss.

Suppose however that he had set odds according to the money wagered – that is, not 3–1 but 2–1 against and 2–1 on respectively. Whichever horse wins, the bookmaker exactly breaks even. The outcome is irrelevant.

In practice the bookmaker sells more than 100% of the race and the odds are shortened to allow for profit (see table). However, the same pattern emerges. Using the actual probabilities can lead to long-term gain but there is always the chance of a substantial short-term loss. For the bookmaker to earn a steady riskless income, he is best advised to assume the horses' probabilities are something different. That done, he is in the surprising position of being disinterested in the outcome of the race, his income being assured.

A note on odds

When a price is quoted in the form n–m against, such as 3–1 against, it means that a successful bet of $\$m$ will be rewarded with $\$n$ plus stake returned. The implied probability of victory (were the price fair) is $m/(m + n)$. Usually the probability is less than half a chance so the first number is larger than the second. Otherwise, what one might write as 1–3 is often called odds of 3–1 on.

Actual probability	25%	75%	
Bets placed	$5000	$10 000	
1. Quoted odds	13–5 against	15–4 on	
Implied probability	28%	79%	Total = 107%
Profit if horse wins	−$3000	$2333	Expected profit = $1000
2. Quoted odds	9–5 against	5–2 on	
Implied probability	36%	71%	Total = 107%
Profit if horse wins	$1000	$1000	Expected profit = $1000

Allowing the bookmaker to make a profit, the odds change slightly. In the first case, the odds relate to the actual probabilities of a horse winning the race. In the second, the odds are derived from the amounts of money wagered.

Chapter 1
Introduction

Financial market instruments can be divided into two distinct species. There are the 'underlying' stocks: shares, bonds, commodities, foreign currencies; and their 'derivatives', claims that promise some payment or delivery in the future contingent on an underlying stock's behaviour. Derivatives can reduce risk – by enabling a player to fix a price for a future transaction now, for example – or they can magnify it. A costless contract agreeing to pay off the difference between a stock and some agreed future price lets both sides ride the risk inherent in owning stock without needing the capital to buy it outright.

In form, one species depends on the other – without the underlying (stock) there could be no future claims – but the connection between the two is sufficiently complex and uncertain for both to trade fiercely in the same market. The apparently random nature of stocks filters through to the claims – they appear random too.

Yet mathematicians have known for a while that to be random is not necessarily to be without some internal structure – put crudely, things are often random in non-random ways. The study of probability and expectation shows one way of coping with randomness and this book will build on probabilistic foundations to find the strongest possible links between claims and their random underlying stocks. The current state of truth is, however, unfortunately complex and there are many false trails through this zoo of the new. Of these, one is particularly tempting.

1.1 Expectation pricing

Consider playing the following game – someone tosses a coin and pays you one dollar for heads and nothing for tails. What price should you pay for this prize? If the coin is fair, then heads and tails are equally likely – about half the time you should win the dollar and the rest of the time you should receive nothing. Over enough plays, then, you expect to make about fifty cents a go. So paying more than fifty cents seems extravagant and less than fifty cents looks extravagant for the person offering the game. Fifty cents, then, seems about right.

Fifty cents is also the expected profit from the game under a more formal, mathematical definition of expectation. A probabilistic analysis of the game would observe that although the outcome of each coin toss is essentially random, this is not inconsistent with a deeper non-random structure to the game. We could posit that there was a fixed measure of likelihood attached to the coin tossing, a *probability* of the coin landing heads or tails of $\frac{1}{2}$. And along with a probability ascription comes the idea of expectation, in this discrete case, the total of each outcome's value weighted by its attached probability. The expected payoff in the game is $\frac{1}{2} \times \$1 + \frac{1}{2} \times \$0 = \$0.50$.

This formal expectation can then be linked to a 'price' for the game via something like the following:

> **Kolmogorov's strong law of large numbers**
> Suppose we have a sequence of independent random numbers X_1, X_2, X_3, and so on, all sampled from the same distribution, which has mean (expectation) μ, and we let S_n be the arithmetical average of the sequence up to the nth term, that is $S_n = (X_1 + X_2 + \ldots + X_n)/n$. Then, with probability one, as n gets larger the value of S_n tends towards the mean μ of the distribution.

If the arithmetical average of outcomes tends towards the mathematical expectation with certainty, then the average profit/loss per game tends towards the mathematical expectation less the price paid to play the game. If this difference is positive, then in the long run it is certain that you will end up in profit. And if it is negative, then you will approach an overall loss with certainty. In the *short term* of course, nothing can be guaranteed, but over time, expectation will out. Fifty cents is a fair price in this sense.

But is it an enforceable price? Suppose someone offered you a play of the game for 40 cents in the dollar, but instead of allowing you a number of plays, gave you just one for an arbitrarily large payoff. The strong law lets you take advantage of them over repeated plays: 40 cents a dollar would then be financial suicide, but it does nothing if you are allowed just one play. Mortgaging your house, selling off all your belongings and taking out loans to the limit of your credit rating would not be a rational way to take advantage of this source of free money.

So the 'market' in this game could trade away from an expectation justified price. Any price might actually be charged for the game in the short term, and the number of 'buyers' or 'sellers' happy with that price might have nothing to do with the mathematical expectation of the game's outcome. But as a guide to a starting price for the game, a ball-park amount to charge, the strong law coupled with expectation seems to have something going for it.

Time value of money

We have ignored one important detail – the time value of money. Our analysis of the coin game was simplified by the payment for and the payoff from the game occurring at the same time. Suppose instead that the coin game took place at the end of a year, but payment to play had to be made at the beginning – in effect we had to find the value of the coin game's contingent payoff not as of the future date of play, but as of now.

If we are in January, then one dollar in December is not worth one dollar now, but something less. Interest rates are the formal acknowledgement of this, and bonds are the market derived from this. We could assume the existence of a market for these future promises, the prices quoted for these bonds being structured, derivable from some interest rate. Specifically:

Time value of money
We assume that for any time T less than some time horizon τ, the value now of a dollar promised at time T is given by $\exp(-rT)$ for some constant $r > 0$. The rate r is then the *continuously compounded* interest rate for this period.

The interest rate market doesn't have to be this simple; r doesn't have to be constant. And indeed in real markets it isn't. But here we assume it is. We can derive a strong-law price for the game played at time T. Paying 50 cents at time T is the same as paying $50\exp(-rT)$ cents now. Why? Because the payment of 50 cents at time T can be guaranteed by buying half a unit of the appropriate bond (that is, promise) now, for cost $50\exp(-rT)$ cents. Thus the strong-law price must be not 50 cents but $50\exp(-rT)$ cents.

Stocks, not coins

What about real stock prices in a real financial market? One widely accepted model holds that stock prices are *log-normally* distributed. As with the time value of money above, we should formalise this belief.

Stock model

We assume the existence of a random variable X, which is normally distributed with mean μ and standard deviation σ, such that the change in the logarithm of the stock price over some time period T is given by X. That is

$$\log S_T = \log S_0 + X, \qquad \text{or} \qquad S_T = S_0 \exp(X).$$

Suppose, now, that we have some claim on this stock, some contract that agrees to pay certain amounts of money in certain situations — just as the coin game did. The oldest and possibly most natural claim on a stock is the *forward*: two parties enter into a contract whereby one agrees to give the other the stock at some agreed point in the future in exchange for an amount agreed now. The stock is being *sold forward*. The 'pricing question' for the forward stock 'game' is: what amount should be written into the contract now to pay for the stock one year in the future?

We can dress this up in formal notation — the stock price at time T is given by S_T, and the forward payment written into the contract is K, thus the value of the contract at its expiry, that is when the stock transfer actually takes place, is $S_T - K$. The time value of money tells us that the value of this claim as of now is $\exp(-rT)(S_T - K)$. The strong law suggests that the expected value of this random amount, $\mathbb{E}\big(\exp(-rT)(S_T - K)\big)$, should be zero. If it is positive or negative, then long-term use of that pricing should lead to one side's profit. Thus one apparently reasonable answer to the pricing question says K should be set so that $\mathbb{E}\big(\exp(-rT)(S_T - K)\big) = 0$, which happens when $K = \mathbb{E}(S_T)$.

What is $\mathbb{E}(S_T)$? We have assumed that $\log(S_T) - \log(S_0)$ is normally distributed with mean μ and variance σ^2 – thus we want to find $\mathbb{E}\big(S_0 \exp(X)\big)$, where X is normally distributed with mean μ and standard deviation σ. For that, we can use a result such as:

> **The law of the unconscious statistician**
> Given a real-valued random variable X with probability density function $f(x)$ then for any integrable real function h, the expectation of $h(X)$ is
> $$\mathbb{E}\big(h(X)\big) = \int_{-\infty}^{\infty} h(x) f(x) \, dx.$$

Since X is normally distributed, the probability density function for X is

$$f(x) = \frac{1}{\sqrt{2\pi\sigma^2}} \exp\left(\frac{-(x-\mu)^2}{2\sigma^2}\right).$$

Integration and the law of the unconscious statistician then tells us that the expected stock price at time T is $S_0 \exp(\mu + \tfrac{1}{2}\sigma^2)$. This is the strong-law justified price for the forward contract; just as with the coin game, it can only be a suggestion as to the market's trading level. But the technique will clearly work for more than just forwards. Many claims are capable of translation into functional form, $h(X)$, and the law of the unconscious statistician should be able to deliver an expected value for them. Discounting this expectation then gives a theoretical value which the strong law tempts us into linking with economic reality.

1.2 Arbitrage pricing

So far, so plausible – but seductive though the strong law is, it is also completely useless. The price we have just determined for the forward could only be the market price by an unfortunate coincidence. With markets where the stock can be bought and sold freely and arbitrary positive and negative amounts of stock can be maintained without cost, trying to trade forward using the strong law would lead to disaster – in most cases there would be *unlimited* interest in selling forward to you at that price.

Why does the strong law fail so badly with forwards? As mentioned above in the context of the coin game, the strong law cannot enforce a price, it only suggests. And in this case, *another completely different mechanism does enforce a price*. The fair price of the contract is $S_0 \exp(rT)$. It doesn't depend on the expected value of the stock, it doesn't even depend on the stock price having some particular distribution. Either counterparty to the contract can in fact *construct* the claim at the *start* of the contract period and then just wait patiently for expiry to exchange as appropriate.

Construction strategy

Consider the seller of the contract, obliged to deliver the stock at time T in exchange for some agreed amount. They could borrow S_0 now, buy the stock with it, put the stock in a drawer and just wait. When the contract expires, they have to pay back the loan – which if the continuously compounded rate is r means paying back $S_0 \exp(rT)$, but they have the stock ready to deliver. If they wrote less than $S_0 \exp(rT)$ into the contract as the amount for forward payment, then they would lose money *with certainty*.

So the forward price is bounded below by $S_0 \exp(rT)$. But of course, the buyer of the contract can run the scheme in reverse, thus writing *more* than $S_0 \exp(rT)$ into the contract would guarantee them a loss. The forward price is bounded above by $S_0 \exp(rT)$ as well.

Thus there is an *enforced* price, not of $S_0 \exp(\mu + \frac{1}{2}\sigma^2)$ but $S_0 \exp(rT)$. Any attempt to strike a different price and offer it into a market would inevitably lead to someone taking advantage of the free money available via the construction procedure. And unlike the coin game, mortgaging the house *would* now be a rational action. This type of market opportunism is old enough to be ennobled with a name – arbitrage. The price of $S_0 \exp(rT)$ is an arbitrage price – it is justified because any other price could lead to unlimited riskless profits for one party. The strong law wasn't wrong – if $S_0 \exp(\mu + \frac{1}{2}\sigma^2)$ is greater than $S_0 \exp(rT)$, then a buyer of a forward contract *expects* to make money. (But then of course, if the stock is expected to grow faster than the riskless interest rate r, so would buyers of the stock itself.) But the existence of an arbitrage price, however surprising, overrides the strong law. To put it simply, if there is an arbitrage price, any other price is too dangerous to quote.

1.3 Expectation vs arbitrage

The strong law and expectation give the wrong price for forwards. But in a certain sense, the forward is a special case. The construction strategy – buying the stock and holding it – certainly wouldn't work for more complex claims. The standard call option which offers the buyer the right but not the obligation to receive the stock for some strike price agreed in advance certainly couldn't be constructed this way. If the stock price ends up above the strike, then the buyer would exercise the option and ask to receive the stock – having it salted away in a drawer would then be useful to the seller. But if the stock price ends up below the strike, the buyer will abandon the option and any stock owned by the seller would have incurred a pointless loss.

Thus maybe a strong-law price would be appropriate for a call option, and until 1973, many people would have agreed. Almost everything appeared safe to price via expectation and the strong law, and only forwards and close relations seemed to have an arbitrage price. Since 1973, however, and the infamous Black–Scholes paper, just how wrong this is has slowly come out. Nowhere in this book will we use the strong law again. Just to muddy the waters, though, *expectation* will be used repeatedly, but it will be as a tool for risk-free *construction*. All derivatives can be built from the underlying – arbitrage lurks everywhere.

Chapter 2
Discrete processes

T he goal of this book is to explore the limits of arbitrage. Bit by bit we will put together a mathematical framework strong enough to be a realistic model of the real financial markets and yet still structured enough to support construction techniques. We have a long way to go, though; it seems wise to start very small.

2.1 The binomial branch model

Something random for the stock and something to represent the time-value of money. At the very least we need these two things – any model without them cannot begin to claim any relation to the real financial market. Consider, then, the simplest possible model with a stock and a bond.

The stock

Just one time-tick – we start at time $t = 0$ and end a short tick later at time $t = \delta t$. We need something to represent the stock, and it had better have some unpredictability, some random component. So we suppose that only two things can happen to the stock in this time: an 'up' move or a 'down' move. With just two things allowed to happen, pictorially we have a branch (figure 2.1).

Our randomness will have some structure – we will assign probabilities to the up and down move: probability p to move up to node 3, and thus $1 - p$

to move down to node 2. The stock will have some value at the start (node 1 as labelled on the picture), call it s_1. This value represents a price at which we can buy and sell the stock in unlimited amounts. We can then hold on to the stock across the time period until time $t = \delta t$. Nothing happens to us in the intervening period by dint of holding on to the stock – there is no charge for holding positive or negative amounts – but at the end of the period it will have a new value. If it moves down, to node 2, then it will have value s_2; up, to node 3, value s_3.

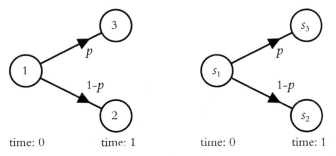

Figure 2.1 The binomial branch

The bond

We also need something to represent the time-value of money – a cash bond. There will be some continuously compounded interest rate r that will hold for the period $t = 0$ to $t = \delta t$ – one dollar at time zero will grow to $\$\exp(r\,\delta t)$. We should be able to lend at that rate, and borrow – and in arbitrary size. To represent this, we introduce a cash bond B which we can buy or sell at time zero for some price, say B_0, and which will be worth a definite $B_0 \exp(r\,\delta t)$ a tick later.

These two instruments are our financial world, and simple though it is it still has uncertainties for investors. Only one of the possible stock values might suit a particular player, their plans surviving or failing by the random outcome. Thus there could be a market for instruments dependent on the value the stock takes at the end of the tick-period. The investor's requirement for compensation based on the future value of the stock could be codified by a function f mapping the two future possibilities, node 2 or node 3, to two rewards or penalties $f(2)$ and $f(3)$. A forward contract, struck at k, for example, could be codified as $f(2) = s_2 - k$, $f(3) = s_3 - k$.

Risk-free construction

The question can now be posed – exactly what class of functions f can be explicitly constructed via a suitable strategy? Clearly the forward can be – as in chapter one, we would buy the stock (cost: s_1), and sell off cash bonds to fund the purchase. At the end of the period, we would be able to hand over the stock and demand $s_1 \exp(r\,\delta t)$ in exchange. The price k of the forward thus has to be $s_1 \exp(r\,\delta t)$ exactly as we would have hoped – priced via arbitrage.

But what about more complex f? Can we still find a construction strategy? Our first guess would be no. The stock takes one of two random values at the end of the tick-period and the value of the derivative would in general be different as well. The probabilities of each outcome for the derivative f are known, thus we also know the expected value of f at the end of the period as well: $(1-p)f(2) + pf(3)$, but we don't know its actual value in advance.

Bond-only strategy

All is not lost, though. Consider a portfolio of just the cash bond. The cash bond will grow by a factor of $\exp(r\,\delta t)$ across the period, thus buying discount bonds to the value of $\exp(-r\,\delta t)[(1-p)f(2) + pf(3)]$ at the start of the period will provide a value equal to $(1-p)f(2) + pf(3)$ at the end. Why would we choose this value as the target to aim for? Because it is the expected value of the derivative at the end of the period – formally:

Expectation for a branch

Let S be a binomial branch process with base value s_1 at time zero, down-value s_2 and up-value s_3. Then the expectation of S at tick-time 1 under the probability of an up-move p is:

$$\mathbb{E}_p(S_1) = (1-p)s_2 + ps_3.$$

Our claim f on S is just as much a random variable as S_1 is – we can meaningfully talk of its expectation. And thus we can meaningfully aim for the expectation of the claim, via the cash bonds. This strategy of construction would at the very least be *expected* to break even. And the value of the starting portfolio of cash bonds might be claimed to be a good predictor of the value of the derivative at the start of the period. The price we would predict for the derivative would be the discounted expectation of its value at the end.

But of course this is just the strong law of chapter one all over again — just thinly disguised as construction. And exactly as before we are missing an element of coercion. We haven't explicitly constructed the two possible values the derivative can take: $f(2)$ and $f(3)$; we have simply aimed between them in a probabilistic sense and hoped for the best.

And we already know that this best isn't good enough for forwards. For a stock that obeys a binomial branch process, its forward price is not suggested by the possible stock values s_2 and s_3, but *enforced* by the interest rate r implied by the cash bond B: namely $s_1 \exp(r\,\delta t)$. The discounted expectation of the claim doesn't work as a pricing tool.

Stocks and bonds together

But can we do any better? Another strategy might occur to us, we have after all two instruments which we can build into a portfolio to hold for the tick–period. We tried using the guaranteed growth of the cash bond as a device for producing a particular desired value, and we chose the expected value of the derivative as our target point. But we have another instrument tied more strongly to the behaviour of both the stock and the derivative than just the cash bond. Namely the stock itself. Suppose we attempted to guarantee not an amount known in advance which we hope will stand as a reasonable predictor for the value of the derivative, but the value of the derivative itself, *whatever it might be*.

Consider a general portfolio (ϕ, ψ), namely ϕ of the stock S (worth ϕs_1) and ψ of the cash bond B (worth ψB_0). If we were to buy this portfolio at time zero, it would cost $\phi s_1 + \psi B_0$.

One tick later, though, it would be worth one of two possible values:

$$\phi s_3 + \psi B_0 \exp(r\,\delta t) \quad \text{after an 'up' move,}$$

$$\text{and} \quad \phi s_2 + \psi B_0 \exp(r\,\delta t) \quad \text{after a 'down' move.}$$

This pair of equations should intrigue us — we have two equations, two possible claim values and two free variables ϕ and ψ. We have two values $f(3)$ and $f(2)$ which we want to duplicate under the appropriate move of the stock, and we have two variables ϕ and ψ which we can adjust. Thus the strategy can reduce to solving the following two simultaneous equations for (ϕ, ψ):

$$\phi s_3 + \psi B_0 \exp(r\, \delta t) = f(3),$$
$$\phi s_2 + \psi B_0 \exp(r\, \delta t) = f(2).$$

Except if perversely s_2 and s_3 are identical − in which case S is a bond not a stock − we have the solutions:

$$\phi = \frac{f(3) - f(2)}{s_3 - s_2},$$

$$\psi = B_0^{-1} \exp(-r\, \delta t) \left(f(3) - \frac{(f(3) - f(2)) s_3}{s_3 - s_2} \right).$$

What can we do with this algebraic result? If we bought this (ϕ, ψ) portfolio and held it, the equations guarantee that we achieve our goal − if the stock moves up, then the portfolio becomes worth $f(3)$; and if the stock moves down, the portfolio becomes worth $f(2)$. We have synthesized the derivative.

The price is right

Our simple model allows a surprisingly prescient strategy. *Any* derivative f can be constructed from an appropriate portfolio of bond and stock. And constructed in advance. This must have some effect on the value of the claim, and of course it does − unlike the expectation derived value, this *is* enforceable in an ideal market as a rational price. Denote by V the value of buying the (ϕ, ψ) portfolio, namely $\phi s_1 + \psi B_0$, which is:

$$V = s_1 \left(\frac{f(3) - f(2)}{s_3 - s_2} \right) + \exp(-r\, \delta t) \left(f(3) - \frac{(f(3) - f(2)) s_3}{s_3 - s_2} \right).$$

Now consider some other market maker offering to buy or sell the derivative for a price P less than V. Anyone could buy the derivative from them in arbitrary quantity, and *sell* the (ϕ, ψ) portfolio to exactly match it. At the end of the tick-period the value of the derivative would exactly cancel the value of the portfolio, *whatever the stock price was* − thus this set of trades carries no risk. But the trades were carried out at a profit of $V - P$ per unit of derivative/portfolio bought − by buying arbitrary amounts, anyone could make arbitrary *risk-free* profits. So P would not have been a rational price for the market maker to quote and the market would quickly have mobilised to take advantage of the 'free' money on offer in arbitrary quantity.

Similarly if a market maker quoted the derivative at a price P greater than V, anyone could sell them it and buy the (ϕ, ψ) portfolio to lock in a risk-free profit of $P - V$ per unit trade. Again the market would take advantage of the opportunity.

Only by quoting a two-way price of V can the market maker avoid handing out risk-free profits to other players – hence V is the only rational price for the derivative at time zero, the start of the tick-period. Our model, though allowing randomness, lets arbitrage creep everywhere – the strong law can be banished completely.

Example – the whole story in one step

We have an interest-free bond and a stock, both initially priced at $1. At the end of the next time interval, the stock is worth either $2 or $0.50. What is the worth of a bet which pays $1 if the stock goes up?

Solution. Let B denote the bond price, S the stock price, and X the payoff of the bet. The picture describes the situation:

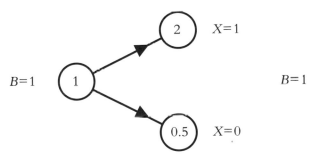

Figure 2.2 Pricing a bet

Buy a portfolio consisting of 2/3 of a unit of stock and a borrowing of 1/3 of a unit of bond. The cost of this portfolio at time zero is $\frac{2}{3} \times$ $1 - \frac{1}{3} \times $1 = 0.33. But after an up-jump, this portfolio becomes worth $\frac{2}{3} \times $2 - \frac{1}{3} \times $1 = 1. After a down-jump, it is worth $\frac{2}{3} \times $0.5 - \frac{1}{3} \times $1 = 0. The portfolio exactly simulates the bet's payoff, so has to be worth exactly the same as the bet. It must be that the portfolio's initial value of $0.33 is also the bet's initial value.

Expectation regained

A surprise still lurks. The strong-law approach may be useless in this model – leaving aside coincidence, expectation pricing involving the probabilities p and $1 - p$ leads to risk-free profits being available. But with an eye to rearranging the equations, we can define a simplifying variable:

$$q = \frac{s_1 \exp(r \, \delta t) - s_2}{s_3 - s_2}.$$

What can we say about q? Without loss of generality, we can assume that s_3 is bigger than s_2. Were q to be less than or equal to 0, then $s_1 \exp(r \, \delta t) \leqslant s_2 < s_3$. But $s_1 \exp(r \, \delta t)$ is the value that would be obtained by buying s_1 worth of the cash bond B at the start of the tick-period. Thus the stock could be bought in arbitrary quantity, financed by selling the appropriate amount of cash bond and a guaranteed risk-free profit made. It is not unreasonable then to eliminate this possibility by *fiat* – specifying the structure of our market to avoid it. So for any *market* in which we have a stock which obeys a binomial branch process S, we have $q > 0$.

Similarly were q to be greater than or equal to 1, then $s_2 < s_3 \leqslant s_1 \exp(r \, \delta t)$ – and this time selling stock and buying cash bonds provides unlimited risk-free gains. Thus the structure of a rational market will force q into $(0, 1)$, the interval of points strictly between 0 and 1 – the same constraint we might demand for a probability.

Now the surprise: when we rewrite the formula for the value V of the (ϕ, ψ) portfolio (try it) we get:

$$V = \exp(-r \, \delta t)\big((1 - q)f(2) + qf(3)\big).$$

Outrageous though it might seem, this is the expectation of the claim under q. This re-appearance of the expectation operator is unsettling.

The price V is *not* the expected future value of the derivative (discounted suitably by the growth of the cash bond) – that would involve p in the above formula. Yet V *is* the discounted *expectation* with respect to *some* number q in $(0, 1)$. If we view the expectation operator as implying some information about the future – a strong-law average over many trials, for example – then V is not what we would unconsciously call the expected value. It sounds pedantic to say it, but V is an expectation, not an expected value. And it is easy enough to check that *this* expectation gives the correct strike for a forward contract: $s_1 \exp(r \, \delta t)$.

Exercise 2.1 Show that a forward contract, struck at k, can be thought of as the payoff f, where $f(2) = s_2 - k$ and $f(3) = s_3 - k$. Now verify, using the formula for V, that the correct strike price is indeed $s_1 \exp(r\,\delta t)$.

2.2 The binomial tree model

From branch to tree. Our single time step was simple to analyse, but it represents a bare minimum as a model. It had a random stock and a cash bond, but it only allowed the stock two possible values at the end of a single time period. Markets are not quite that straightforward. But if we could build the branch model up into something more sophisticated, then we could transfer its results into a larger, better model. This is the intention of this section – we shall build a tree out of branches, and see what survives.

Our financial world will again be just two instruments – a discount bond B and a stock S. Unlimited amounts of either can be bought and sold without transaction costs, default risks, or bid-offer spreads. But now, instead of a single time-period, we will allow many, stringing the individual δts together.

The stock

Changes in the value of the stock S must be random – the market demands that – but the randomness can have structure. Our mini-stock from the binomial branch model allowed the stock to change to just two values at the end of the time period, and we shall keep that structure. But now, we will string these choices together into a tree. The very first time period, from $t = 0$ to $t = \delta t$, will be just as before (a tree of branches starts with just one simple branch). If the value of S at time zero is $S_0 = s_1$, then the *actual* value one tick later is not known but the range of possibilities *is* – S_1 has only two possible values: s_2 and s_3.

Now, we must extend the branch idea in a natural fashion. One tick δt later still, the stock again has two possibilities, *but dependent on the value at tick-time 1*; hence there are four possibilities. From s_2, S_2 can be either s_4 or s_5; from s_3, S_2 can be either s_6 or s_7.

17

As the picture suggests, at tick–time i, the stock can have one of 2^i possible values, though of course *given* the value at tick–time $(i-1)$, there are still only two admissible possibilities: from node j the process either goes down to node $2j$ or up to node $2j+1$.

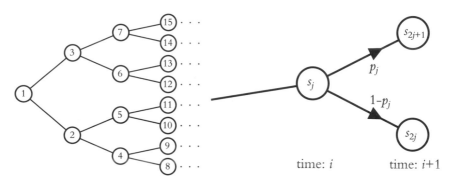

Binary tree with numbered nodes   Stock price development

Figure 2.3

This tree arrangement gives us considerably more flexibility. A claim can now call on not just two possibilities, but any number. If we think that a thousand random possible values for a stock is a suitable level of complexity, then we merely have to set δt small enough that we get ten or so layers of the tree in before the claim time t. We also have a richer allowed structure of probability. *Each* up/down choice will have an attached probability of it being made. From the standpoint of notation, we can represent this pair of probabilities (which must sum to 1) by just one of them (the up probability) p_j, the probability of the stock achieving value s_{2j+1}, given its previous value of s_j. The probability of the stock moving down, and achieving value s_{2j}, is then $1 - p_j$. Again this is shown in the picture.

The cash bond

To go with our grown–up stock, we need a grown–up cash bond. In the simple branch model, the cash bond behaved entirely predictably; there was a known interest rate r which applied across the period making the cash bond price increase by a factor of $\exp(r\,\delta t)$. There is no reason to impose such a strict condition – we don't have to have a constant interest rate known for the entire tree in advance but instead we could have a sequence of interest rates, $R_0, R_1, \ldots$, each known at the *start* of the appropriate tick period. The value of the cash bond at time $n\,\delta t$ thus be $B_0 \exp(\sum_{i=0}^{n-1} R_i\,\delta t)$.

It is worth contrasting the cash bond and the stock. We have admitted the possibility of randomness in the cash bond's behaviour (though in fact we will not yet be particularly interested in its exact form). But compared to the stock it is a very different sort of randomness. The cash bond B has the same structure as the time value of money. The interest which must be paid or earned on cash can change over time, but the value of a cash holding at the *next* tick point *is* always known, because it depends only on the interest rate already known at the *start* of the period.

But for simplicity's sake, we will now keep a constant interest rate r applying everywhere in the tree, and in this case the price of the cash bond at time $n\,\delta t$ is $B_0 \exp(rn\,\delta t)$.

Trees are complex

At this stage, the binomial structure of the tree may seem rather arbitrary, or indeed unnecessarily simplistic. A tree is better than a single branch, but it still won't allow continuous fluid changes in stock and bond values. In fact, as we shall see, it more than suits our purpose. Our final goal, an understanding of the limitations (or lack of them) of risk-free construction when the underlying stocks take continuous values in continuous time, will draw directly and naturally on this starting point. And as δt tends to zero, this model will in fact be more than capable of matching the models we have in mind. Perhaps more pertinently, before we abandon the tree as simplistic, we had better check that it hasn't become too complex for us to make any analytic progress at all.

Backwards induction

In fact most of the hard work has already been done when we examined the branch model. Extending the results and intuitions of section 2.1 to an entire binomial tree is surprisingly straightforward. The key idea is that of backwards induction – extending the construction portfolio back one tick at a time from the claim to the required starting place.

Consider, then, a general claim for our stock S. When we examined a single branching of our tree, we had the function f dependent only on the node chosen at the end of a single tick period – here we can extend the idea of a claim to cover not only the value of S at the time the claim is exercised but also the history of S up until that point.

The tree structure of the stock was not entirely arbitrary – it embodies a one-to-one relationship between a node and the history of the stock's path up to and including that node. No other history reaches that node; and trivially no other node is reached by that history. This is precisely that condition that allows us actually to associate a claim value with a particular end-node on our tree. We shall also insist on the finiteness of our tree. There must be some final tick-time at which the claim is fully determined. A condition not unreasonable in the real financial world. A general claim can be thought of as some function on the nodes at this claim time-horizon.

The two-step

We know that the expectation operator can be made to work for a single branch – here, then, we must wade through the algebra for two time-steps, three branches stuck together into a tree. If two time-steps work, then so will many.

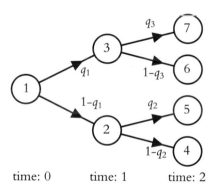

time: 0   time: 1   time: 2

Figure 2.4 Double fork at time 0

Suppose that the interest rate over any branch is constant at rate r. Then there exists some set of suitable q_js such that the value of the derivative at node j at tick-time i, $f(j)$, is

$$f(j) = e^{-r\,\delta t}\big(q_j f(2j+1) + (1-q_j)f(2j)\big).$$

That is the discounted expectation under q_j of the time-$(i+1)$ claim values $f(2j+1)$ and $f(2j)$. So in our two-step tree (figure 2.4), the two forks from node 3 to nodes 6 and 7, and from node 2 to nodes 4 and 5, are both structurally identical to the simple one-step branch. This means that $f(3)$ comes from $f(6)$ and $f(7)$ via

$$f(3) = e^{-r\,\delta t}\big(q_3 f(7) + (1 - q_3)f(6)\big),$$

and similarly, $f(2)$ comes from $f(4)$ and $f(5)$, with

$$f(2) = e^{-r\,\delta t}\big(q_2 f(5) + (1 - q_2)f(4)\big).$$

Here q_j is the probability $\big(s_j \exp(r\,\delta t) - s_{2j}\big)/(s_{2j+1} - s_{2j})$, so for instance

$$q_2 = \frac{s_2 \exp(r\,\delta t) - s_4}{s_5 - s_4}, \quad \text{and} \quad q_3 = \frac{s_3 \exp(r\,\delta t) - s_6}{s_7 - s_6}.$$

But now we have a value for the claim at time 1; it is worth $f(3)$ if the first jump was up, and $f(2)$ if it was down. But this initial fork from node 1 to nodes 2 and 3 also has the single branch structure. Its value at time zero must be

$$f(1) = e^{-r\,\delta t}\big(q_1 f(3) + (1 - q_1)f(2)\big).$$

Thus the value of the claim at time zero has the daunting looking expression formed by combining the three equations above,

$$f(1) = e^{-2r\,\delta t}\Big(q_1 q_3 f(7) + q_1(1 - q_3)f(6)$$
$$+ (1 - q_1)q_2 f(5) + (1 - q_1)(1 - q_2)f(4)\Big).$$

We haven't formally defined expectation on our tree, but it is clear what it must be.

Path probabilities

The probability that the process follows a particular path through the tree is just the product of the probabilities of each branch taken. For example, in figure 2.4, the chance of going up twice is the product $q_1 q_3$, the chance of going up and then down is $q_1(1 - q_3)$, and so on.

This is a case of the more general slogan that when working with independent events, the probabilities multiply.

Expectation on a tree

The expectation of some claim on the final nodes of a tree is the sum over those nodes of the claim value weighted by the probabilities of paths reaching it.

A two-step tree has four possible paths to the end. But each path carries two probabilities attached to it, one for the first time step and one for the second, thus the path-probability, the probability of following any particular path, must be the product of these.

The expectation of a claim is then the total of the four outcomes each weighted by this path-probability. But examine the expression we have derived above – it is of course precisely the expectation of the claims $f(7)$, $\ldots$, $f(4)$, discounted by the appropriate interest-rate factor $e^{-2r\,\delta t}$, under the probabilities $q_1 q_3$, $q_1(1-q_3)$, $(1-q_1)q_2$, $(1-q_1)(1-q_2)$ corresponding to the 'probability tree' (q_1, q_2, q_3).

For claim pricing and expectation, a two-step tree *is* simply three branches. And so on.

The inductive step

Returning to our general tree over n periods, we start at its final layer. All nodes here have claim values and are in pairs, the ends of single branchings. Consider any one of these final branchings, from a node at time $(n-1)$ to two nodes at time n. The results from section 2.1 provide a risk-free construction portfolio (ϕ, ψ) of stock and bond at the root of the branch that can generate the time n claim amount. (Both our grown-up stock and the cash bond are indistinguishable *over a single branching* from the stock and bond of the simple model.)

Thus the nodes at time $(n-1)$ are all roots of branches that end on the claim layer and have arbitrage guaranteed values for the derivative attached – claim-values in their own right now insisted on not by the investor's contract (that only applies to the final layer) but by arbitrage considerations. Thus we can work back from enforced claims at the final layer to equally strongly enforced claim-values at the layer before. *This* is the inductive step – we have moved the claims on the final layer back one step.

The inductive result

By repeating the inductive step, we will sweep backwards through the tree. Each layer will fix the value of the derivative on the layer before, because each layer is only separated from the layer before by simple branches. What we have done is essentially a recursive filling in process. The investor filled in the nodes at the end of the tree with claims – we filled in the rest by constructing

(ϕ, ψ) portfolios at each branching which guaranteed the correct outcome at the next step.

We will reach the root of the entire tree with a single value. This is the time–zero value of the final derivative claim – why? Because just as for the single branch, there is a construction portfolio which, *though it will change at each tick time*, will inexorably lead us to the claim payoff required, *whatever* path the stock actually takes.

We now have some idea of the complexity of the construction portfolios that will be required. Instead of a single amount of stock ϕ, we now have a whole number of them, one per node. And as fate casts the die and the stock jumps on the tree, so this amount will jump as well. Perverse though it may seem for a guaranteed construction procedure, the construction portfolios (ϕ_i, ψ_i) are *also* random, just like the stock. But there is a vital structural difference – they are known *just in time* to be useful, unlike the stock value they are known one-step in advance.

Arbitrage has worked its way into the tree model as well. The fact that the tree is simply lots of branches was enough to banish the strong law here as well. All claims can be constructed from a stock and bond portfolio, and thus all claims have an arbitrage price.

Expectation again

The strong law may be useless, but what about expectation? We had no need of the probabilities p_j, but the re-emergence of the expectation operator is not just a coincidence peculiar to the simplicities of the branch model. Yet again the expectation operator will appear with the correct result – just as the conclusion from the previous section was that with respect to a suitable 'probability', the expectation operator provided the correct *local* hedge, here we will see that the expectation operator with respect to some suitable *set* of 'probabilities' also provides the correct *global* structure for a hedge.

A worked example

We can give a concrete demonstration of how this works. The tree in Figure 2.5 is called *recombinant* as different branches can come back together, or recombine, at the same node. Such trees are computationally much easier to work with, as long as we remember that there is more than one path to the final nodes. The tree nodes are the stock prices, s, and at each node the process will go up with probability 3/4 and down with probability 1/4. (For simplicity, interest rates are zero.)

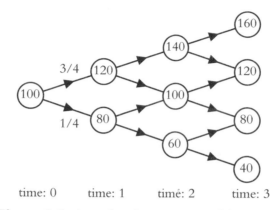

Figure 2.5 A stock price on a recombinant tree

What is the value of an option to buy the stock for 100 at time 3?

It is easy to fill in the value of the claim on the time 3 column. Reading from top to bottom, the claim has values then of 60, 20, 0 and 0.

We shall now need our equations for the new probabilities q and the claim values f. As the interest rate r is zero, these equations are a little simpler. If we are about to move either 'up' or 'down', then the (risk–neutral) probability q is

$$q = \frac{s_{now} - s_{down}}{s_{up} - s_{down}}$$

and the value of a claim, f, now is

$$f_{now} = q f_{up} + (1 - q) f_{down}.$$

We calculate that the new q-probabilities are exactly 1/2 at each and every node. Now we can work out the value of the option at the penultimate time 2 by applying the up–down formulae to the final nodes in adjacent pairs. Figure 2.6 shows the result of the first two such calculations.

We can complete filling in the nodes on level 2, and then repeat the process on level 1, and so on. At the end of this process we have the completed tree (figure 2.7).

The price of the option at time zero is 15. We can trace through our hedge, using the formula that, at any current time, we should hedge

$$\phi = \frac{f_{\text{up}} - f_{\text{down}}}{s_{\text{up}} - s_{\text{down}}}$$

units of stock.

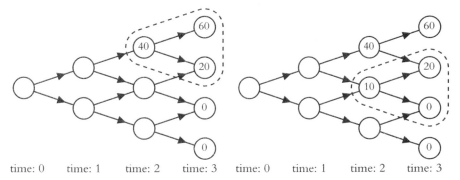

Figure 2.6 The option claims and claim-values at time 2

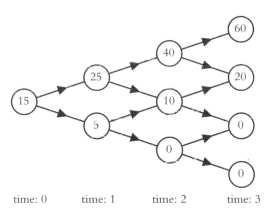

Figure 2.7 The option claim tree

Time 0 We are given 15 for the option. We calculate ϕ as $(25 - 5)/(120 - 80) = 0.5$. Buying 0.5 units of stock costs 50, so we need to borrow an additional 35.

Suppose the stock now goes up to 120

Time 1 The new ϕ is $(40 - 10)/(140 - 100) = 0.75$, so we buy another 0.25

units of stock at its new price, taking our total borrowing to 65.

Suppose the stock goes up again to 140

Time 2 The new ϕ is $(60 - 20)/(160 - 120) = 1$, so we take our stock holding up to 1, making our debt now 100.

Finally suppose the stock goes down to 120

Time 3 The option will be in the money, and we are exactly placed to hand over one unit of stock and receive 100 in cash to cancel our debt. (In fact, the same would have happened if the stock had gone up to 160 instead.)

The table below shows exactly how the various processes change over time. The portfolio strategies shown are those in force for the previous tick–period, for instance, ϕ_1 units of stock are held during the interval from $i = 0$ to $i = 1$. The option value matches the worth of both the old and the new portfolios, for instance V_1 equals both $\phi_1 S_1 + \psi_1$ and $\phi_2 S_1 + \psi_2$.

Table 2.1 Option and portfolio development

Time i	Last Jump	Stock Price S_i	Option Value V_i	Stock Holding ϕ_i	Bond Holding ψ_i
0	–	100	15	–	–
1	up	120	25	0.50	−35
2	up	140	40	0.75	−65
3	down	120	20	1.00	−100

This was the rosy scenario. What would have happened if the initial jump had been down?

Suppose the stock goes down to 80

Time 1 This time, the new ϕ is $(10 - 0)/(100 - 60) = 0.25$. We sell half our stock holding and reduce our debt to 15.

Suppose now the stock goes up to 100

Time 2 The next hedge is $(20 - 0)/(120 - 80) = 0.50$. We buy an extra 0.25 units of stock and our borrowing mounts to 40.

Suppose the stock goes down again to 80

Time 3 Our stock is now worth 40, exactly cancelling the debt. But the option is out of the money, so overall we have broken even.

Table 2.2 Option and portfolio development along a different path

Time i	Last Jump	Stock Price S_i	Option Value V_i	Stock Holding ϕ_i	Bond Holding ψ_i
0	–	100	15	–	–
1	down	80	5	0.50	−35
2	up	100	10	0.25	−15
3	down	80	0	0.50	−40

We note that all the process above (S, V, ϕ and ψ) depend on the sequence of up-down jumps. In particular, ϕ and ψ are random too, but depend only on the jumps made up to the time when you need to work them out.

 Exercise 2.2 Repeat the above calculations for a digital contract which pays off 100 if the stock ends higher than it started.

The expectation result is still here. Under the probability q, the chances of each of the final nodes are (running from top to bottom) 1/8, 3/8, 3/8, and 1/8. The expectation of the claim is indeed 15 under these probabilities, but certainly not under the model probabilities of 3/4-up and 1/4-down. (That gives node probabilities of 27/64, 27/64, 9/64 and 1/64, and a claim expectation of 33.75.)

Conclusions

We can sum up. The tree structure ensured that any claim provides just *one* possible value for its implied derivative instrument at every node or else arbitrage intervened. Claim led to claim-value led to claim-value via backwards induction until the entire tree was filled in. Arbitrage spreads into every branch and thus across any tree.

Something else happened as well – each branchlet carries its own probability q_j under which fixing the value at the branchlet's root can be given by a local expectation operator with parameter q_j. The cost of the local construction portfolio (ϕ_j, ψ_j) can be written as a discounted expectation. But a string of local construction portfolios is a global construction *strategy* guaranteeing a value. Thus the global discounted expectation operator gives the value of claims on a tree as well.

Summary

$$q = \frac{e^{r\,\delta t} s_{\text{now}} - s_{\text{down}}}{s_{\text{up}} - s_{\text{down}}}$$

$$\phi = \frac{f_{\text{up}} - f_{\text{down}}}{s_{\text{up}} - s_{\text{down}}}$$

$$f_{\text{now}} = e^{-r\,\delta t}\left(q f_{\text{up}} + (1-q) f_{\text{down}}\right)$$

$$V = f(1) = \mathbb{E}_{\mathbb{Q}}(B_T^{-1} X)$$

$$\psi = B_{\text{now}}^{-1}\left(f_{\text{now}} - \phi s_{\text{now}}\right)$$

where

q : arbitrage probability of up-jump r : interest rate in force over period
f : claim value time-process s : stock price process
ϕ : stock holding strategy B : bond price process, $B_0 = 1$
ψ : bond holding strategy $\mathbb{Q}$: measure made up of the qs
V : claim value at time zero X : claim payoff
δt : length of period T : time of claim payoff

2.3 Binomial representation theorem

The expectation operator is much more general and constructive an operator than its conventional probabilistic role suggests. We can raise the apparently coincidental finding that there exists some set of q_j under which any derivative can be priced by a numerically trivial discounted expectation operation to the status of a theorem. Though it seems strangely formal here where we have the comfort of a pictured tree, when we move to continuous models we shall be glad of any guidance – in the continuous case intuition will often fail. And far from vanishing, the expectation result carries across to the continuous model with ease.

It is in this spirit, then, that we derive the binomial representation theorem.

Illustrated definitions

We must start with some formal definitions of concepts we have, in many cases, already met informally. There are seven separate definitions and each will be illustrated by an example on the double forked tree with seven nodes (figure 2.8).

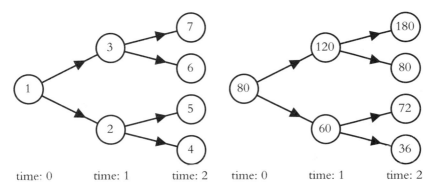

time: 0	time: 1	time: 2

Figure 2.8 Tree with node numbers   **Figure 2.9** Tree with price process

(i) We will call the set of possible stock values, one for each node of the tree, and their pattern of interconnections, a *process S*. One possible process S on our tree is shown in figure 2.9. The random variable S_i denotes the value of the process at time i, for instance S_1 is either 60 or 120 depending on whether we are at node 2 or node 3.

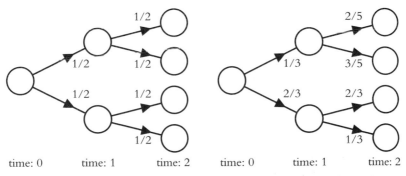

Figure 2.10a The measure $\mathbb{P}$ **Figure 2.10b** The measure $\mathbb{Q}$

(ii) Separate from the process S, we will call the set of 'probabilities' (p_j)

or (q_j) a *measure* $\mathbb{P}$ or $\mathbb{Q}$ on the tree. The measure describes how likely any up/down jump is at each node, represented by p_j, the probability of moving upwards from node j. We could choose a simple measure $\mathbb{P}$ with all jumps equally likely (figure 2.10a) or a more complex measure like $\mathbb{Q}$ (shown in figure 2.10b).

Notice that in our formal system, we have separated two components that would normally be seen as intimately connected parts of the same whole – the probability of an up-move, and where the up-move is to. They may not seem too different in character but the lesson of both the preceding sections is that this intuitive elision is unwise. We didn't need the real world measure $\mathbb{P}$ in order to find the measure which allowed risk-free construction. That measure was a function of S and no function of $\mathbb{P}$. The size and interrelation of up-moves affects the values of derivatives, the probabilities of achieving them does not.

This separation of process and measure isn't artificial – it is fundamental to everything we have to do. Put crudely, the strong law failed precisely because it paid attention to both S and $\mathbb{P}$, not S alone.

(iii) A *filtration* $(\mathcal{F}_i)$ is the history of the stock up until tick–time i on the tree. The filtration starts at time zero with $\mathcal{F}_0$ equal to the path consisting of the single node 1, that is $\mathcal{F}_0 = \{1\}$. By time 1, the filtration will either be $\mathcal{F}_1 = \{1, 2\}$ if the first jump was down, or $\mathcal{F}_1 = \{1, 3\}$ if it was up. In full the filtration associated with each node is

Table 2.3 The filtration process

node	1	2	3	4	5	6	7
filtration	$\{1\}$	$\{1,2\}$	$\{1,3\}$	$\{1,2,4\}$	$\{1,2,5\}$	$\{1,3,6\}$	$\{1,3,7\}$

It thus corresponds to a particular node achieved at time i. Why? Because the binomial structure ensures it – check for yourself that there is only one path to any given node. The filtration fixes a history of choices, and thus fixes a node. To know where you are is the same as knowing the filtration (at least in non-recombinant trees).

(iv) A *claim* X on the tree is a function of the nodes at a claim time-horizon T. Or equivalently it is a function of the filtration $\mathcal{F}_T$, thanks to the one-to-one relationship between nodes and paths. For instance, the value of the process at time 2, S_2, is a claim, as is the value of a call struck at 70 and the maximum price the stock attained along its path (table 2.4).

Table 2.4 Some claims at time 2

time 2 node	S_2	$(S_2 - 70)^+$	$\max\{S_0, S_1, S_2\}$
7	180	110	180
6	80	10	120
5	72	2	80
4	36	0	80

The crucial difference between a claim and a process, is that the claim is only defined on the nodes at time T, while a process is defined at all times up to and including T.

Table 2.5 Conditional expectation against filtration value

Expectation	Filtration value	Value
$\mathbb{E}_\mathbb{P}(S_2\|\mathcal{F}_0)$	$\{1\}$	$(180 + 80 + 72 + 36)/4 = 92$
$\mathbb{E}_\mathbb{P}(S_2\|\mathcal{F}_1)$ ·	$\{1,3\}$	$\frac{1}{2}(180 + 80) = 130$
	$\{1,2\}$	$\frac{1}{2}(72 + 36) = 54$
$\mathbb{E}_\mathbb{P}(S_2\|\mathcal{F}_2)$	$\{1,3,7\}$	180
	$\{1,3,6\}$	80
	$\{1,2,5\}$	72
	$\{1,2,4\}$	36

(v) The conditional *expectation operator* $\mathbb{E}_\mathbb{Q}(\cdot|\mathcal{F}_i)$ extends our idea of expectation to two parameters – a measure $\mathbb{Q}$ and a history $\mathcal{F}_i$. The measure $\mathbb{Q}$ we might have guessed – it tells us which 'probabilities' to use in determining path-probability and thus the expectation. But so far we have only been interested in taking expectations along the whole of a path from time zero, and it is useful to take expectations from later starting points. The filtration serves this purpose. For a claim X, the quantity $\mathbb{E}_\mathbb{Q}(X|\mathcal{F}_i)$ is the expectation of X along the latter portion of paths which have initial segment $\mathcal{F}_i$. We regard the node reached at time i as the new root of our tree, and take expectations of future claims from there. This conditional expectation has an enforced dependence on the value of the filtration $\mathcal{F}_i$, and so is itself a random variable.

For each node at time i, $\mathbb{E}_{\mathbb{Q}}(X|\mathcal{F}_i)$ is the expectation of X *if we have already got to that node*. As an example, we take $\mathbb{P}$ to be the measure in figure 2.10a and X to be the claim S_2 (table 2.5).

Sensibly enough, starting at the root gives the same answer as the unconditioned expectation $\mathbb{E}_{\mathbb{P}}(S_2)$, whereas 'starting' at time 2 leaves no further time for development, so $\mathbb{E}_{\mathbb{P}}(S_2|\mathcal{F}_2) = S_2$, for every possible value of the filtration $\mathcal{F}_2$.

We could also see $\mathbb{E}_{\mathbb{P}}(X|\mathcal{F}_i)$ as a process in i. In the case of $X = S_2$, it is shown in figure 2.11. In this way we can convert a claim into a process, given a measure.

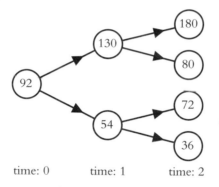

time: 0   time: 1   time: 2

Figure 2.11 Conditional expectation process $\mathbb{E}_{\mathbb{P}}(S_2|\mathcal{F}_i)$

(vi) A *previsible* process $\phi = \phi_i$ is a process on the *same* tree whose value at any given node at time-tick i is dependent only on the history up to one time-tick earlier, $\mathcal{F}_{i-1}$. What can we say about a previsible process? Given the one-to-one relationship between nodes and histories on our binary tree, it is certainly a binomial tree process in its own right, whose values are well defined at each node later than time zero. But *compared to the main process S, it is known one node in advance*. It doesn't seem to notice branches until one time-step after they have happened. For instance a random bond price process B_i would be previsible, as is the delayed price process $\phi_i = S_{i-1}$, $i \geqslant 1$ (figure 2.12). It is not always sensible to define the value that a previsible process has at time zero.

Previsible processes will play the part of trading strategies, where we cannot tell in advance where prices are going to go. This is an essential feature of any model that excludes arbitrage (or insider trading).

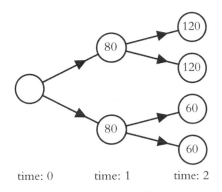

time: 0   time: 1   time: 2

Figure 2.12 The previsible process S_{i-1}

Our final definition is probably the most important of all — one question that we must surely ask soon is: what *is* the risk-free construction measure? Is it specific to the task in hand, or is it special in some other way as well?

(vii) A process S is a *martingale* with respect to a measure $\mathbb{P}$ and a filtration $(\mathcal{F}_i)$ if

$$\mathbb{E}_{\mathbb{P}}(S_j|\mathcal{F}_i) = S_i, \qquad \text{for all } i \leqslant j.$$

This daunting expression needs expansion. Written out, for S to be a martingale with respect to a measure $\mathbb{P}$, it means that the future expected value at time j of the process S *under measure* $\mathbb{P}$ (for of course our formal expectation demands a measure, it has no meaning without one) conditional on its history up until time i is merely the process' value at time i.

Re-written again, that means the process S has no drift under $\mathbb{P}$, no bias up or down in its value under the expectation operator $\mathbb{E}_{\mathbb{P}}$. If the process has value 100 at some point, then its conditional expected value under $\mathbb{P}$ is 100 thereafter.

Example (1). The process which constantly takes a fixed value is, rather trivially, a martingale with respect to all possible measures.

Example (2). Our illustrative process S is actually a martingale under the measure $\mathbb{Q}$ given in figure 2.10b. For instance $\mathbb{E}_{\mathbb{Q}}(S_1|\mathcal{F}_0)$ equals

$\frac{1}{3} \times 120 + \frac{2}{3} \times 60 = 80$, and 80 is indeed the value of S_0. Slightly harder, $\mathbb{E}_\mathbb{Q}(S_2|\mathcal{F}_1)$ equals $\frac{2}{5} \times 180 + \frac{3}{5} \times 80 = 120$ if the first jump was up, which matches the value S_1 takes if the first jump is up. The down–jump case and all the others need to be checked separately.

Example (3). The conditional expectation process $N_i = \mathbb{E}_\mathbb{P}(S_2|\mathcal{F}_i)$ is a $\mathbb{P}$-martingale. Because of the nature of its definition we only need to check that $\mathbb{E}_\mathbb{P}(N_1|\mathcal{F}_0)$ is equal to N_0. As this is just $\frac{1}{2} \times 130 + \frac{1}{2} \times 54 = 92$, it is immediate.

The last example above is a particular example of a general result.

> **The conditional expectation process of a claim**
> For any claim X, the process $\mathbb{E}_\mathbb{P}(X|\mathcal{F}_i)$ is always a $\mathbb{P}$-martingale.

To see this to be true, we need to use the fact that

$$\mathbb{E}_\mathbb{P}\left(\mathbb{E}_\mathbb{P}(X|\mathcal{F}_j)\,\middle|\,\mathcal{F}_i\right) = \mathbb{E}_\mathbb{P}(X|\mathcal{F}_i), \qquad i \leqslant j.$$

In other words, that conditioning firstly on the history up to time j and then conditioning on the history up to an earlier time i is the same as just conditioning originally up to time i. This result is called the *tower law*.

Given the tower law, an easy check of whether a process is a $\mathbb{P}$-martingale or not is to compare the process S_i itself with the conditional expectation process of its terminal value $\mathbb{E}_\mathbb{P}(S_T|\mathcal{F}_i)$. Only if these are identical is the process a $\mathbb{P}$-martingale.

We must also take the $\mathbb{P}$ dependence seriously. The process S is not a martingale on its own, it is a $\mathbb{P}$-martingale, it is a martingale with respect to the measure $\mathbb{P}$. And of course, exactly the same process can be a martingale with respect to one measure and not to another. For instance, our illustrative process S is not a $\mathbb{P}$-martingale (because figure 2.9 and figure 2.11 are different), but it is a $\mathbb{Q}$-martingale, where $\mathbb{Q}$ is given in figure 2.10b. Such a $\mathbb{Q}$ is called a *martingale measure* for S.

 Exercise 2.3 Check that $\mathbb{E}_\mathbb{Q}(S_2|\mathcal{F}_i)$ is the same as S_i, and so prove that S is a $\mathbb{Q}$-martingale.

Binomial representation theorem

We can now write down our theorem.

> **Binomial representation theorem**
> Suppose the measure $\mathbb{Q}$ is such that the binomial price process S is a $\mathbb{Q}$-martingale. If N is any other $\mathbb{Q}$-martingale, then there exists a previsible process ϕ such that
>
> $$N_i = N_0 + \sum_{k=1}^{i} \phi_k \, \Delta S_k,$$
>
> where $\Delta S_i := S_i - S_{i-1}$ is the change in S from tick-time $i - 1$ to i, and ϕ_i is the value of ϕ at the appropriate node at tick-time i.

We can get from N_0 to N_i previsibly, with steps we know in advance. The proof is formal but straightforward – with the work we have put in already, this kind of manipulation should be second nature.

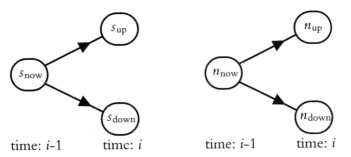

time: i-1   time: i time: i-1   time: i

Figure 2.13 The branch geometry (process S on left; process N on right)

Consider a single branching from a node at tick-time $i - 1$ to two nodes 'up' and 'down' at tick-time i. The structure of the tree ensures that the history $\mathcal{F}_i$ has two choices beyond $\mathcal{F}_{i-1}$, corresponding to the up jump and down jumps respectively. The increments over the branch of the processes S and N are

$$\Delta S_i = S_i - S_{i-1} \quad \text{and} \quad \Delta N_i = N_i - N_{i-1}.$$

The variability that these increments contain depends on the geometry of the branch itself (figure 2.13).

There are only two places to go, so any random variable dependent on the branch is fully determined by its width size and a constant offset depending only on $\mathcal{F}_{i-1}$. So if we want to construct one random process out of another, it will in general be a construction based on a scaling (to match the widths) and a shift (to match the offsets).

Consider then the scaling first. The size of the difference between the up and down jump values is $\delta s_i = s_{\text{up}} - s_{\text{down}}$ for S and $\delta n_i = n_{\text{up}} - n_{\text{down}}$ for N, both of these dependent only on the filtration $\mathcal{F}_{i-1}$. So we define ϕ_i to be the ratio of these branch widths, that is

$$\phi_i = \frac{\delta n_i}{\delta s_i}.$$

Now we can worry about the shift – the N-increment ΔN_i must be given by the scaled increment $\phi_i \Delta S_i$ plus an offset k, this k again determined only by $\mathcal{F}_{i-1}$. That is

$$\Delta N_i = \phi_i \, \Delta S_i + k, \qquad \text{for } \phi_i \text{ and } k \text{ known by } \mathcal{F}_{i-1}.$$

But S and N are $\mathbb{Q}$-martingales, that is $\mathbb{E}_{\mathbb{Q}}(\Delta N_i | \mathcal{F}_{i-1})$ and $\mathbb{E}_{\mathbb{Q}}(\Delta S_i | \mathcal{F}_{i-1})$ are both zero – the increments have zero expectation conditional on the history $\mathcal{F}_{i-1}$. The scaling factor ϕ_i is previsible, that is known by time $i - 1$, so we also have $\mathbb{E}_{\mathbb{Q}}(\phi_i \, \Delta S_i | \mathcal{F}_{i-1}) = 0$. Thus the offset k must be zero as well $(0 = 0 + k)$.

So the general scale and shift reduces in the case where S and N are both $\mathbb{Q}$-martingales to just a scaling

$$\Delta N_i = \phi_i \, \Delta S_i.$$

And of course induction ties all these increments together to give the result we want.

Financial application

We now have a theorem; but it is a formal theorem about binomial tree processes and measures. Nowhere in our proof do we consider portfolios of a stock and bond; nowhere do we consider arbitrage or market implications. We go through many of the same steps as we had to in section 2.2, but we haven't reached a financial conclusion. How then can we use the binomial representation theorem for pricing?

In our binomial tree model for the market, the stock follows a binomial process S. And if there were a measure $\mathbb{Q}$ which made S a martingale, we could use the representation theorem to represent some other martingale N_i in terms of the stock price. The previsible ϕ from the theorem could act as a construction strategy. At each point we could buy the appropriate ϕ_i of the stock and we would follow the gains and losses of the martingale N_i.

We would be able to match the martingale step for step, starting where it starts and finishing where it ends, wherever that might be. If the martingale ended in a claim, than that claim would have been synthesized.

Two things stand in our way, though. Firstly we have a claim X, not a martingale. And though we would like to end up at the claim, the claim doesn't start or end anywhere. It isn't a process, it's a random variable. Secondly, we have not just a stock but a cash bond as well. X-ray vision or intuition would suggest that the ϕ_i of the binomial representation theorem is going to be a vital part of our formal construction strategy but, to use the notation of earlier, we need a ψ_i as well. With each stock holding comes a bond holding.

First things first. The claim X is a random variable but we have already seen one trick for turning random variables into processes. Given any measure $\mathbb{Q}$, we can form the process

$$E_i = \mathbb{E}_{\mathbb{Q}}(X|\mathcal{F}_i),$$

by taking conditional expectations. Even better, as we have already observed, whatever measure $\mathbb{Q}$ we choose, E_i is automatically a $\mathbb{Q}$-martingale. Thus if we find $\mathbb{Q}$, a measure under which S_i is a $\mathbb{Q}$-martingale, the appropriate E_i is one as well.

What about the cash bond? Ultimately we will simply have to grind through the algebra but a bit of intuition can guide us to what the answer might look like. The cash bond B_i represents the growth of money – $1 today is not the same as $1 at time i, all things being equal. One dollar today is like B_i dollars at time i. But we would like to be in a world without the growth of money – so we could simply factor it away.

The bond process B_i is previsible and positive. We can assume without loss of generality that $B_0 = 1$.

(i)   The process B_i^{-1} is another previsible process, just like B_i itself. Call this the *discount process*.

(ii)   Define $Z_i := B_i^{-1} S_i$ which is just as well a defined process as S itself and it subsists on the same binomial tree. Call this the *discounted stock process*.

(iii)   The value $B_T^{-1} X$ is also a claim and because of the simple mapping from Z to S it's as much a claim on Z as S. Call this the *discounted claim*.

What, then, now?

Construction strategy

Let's try out the trick. With $\mathbb{Q}$ such that Z is a $\mathbb{Q}$-martingale and claim X, there is a $\mathbb{Q}$-martingale process produced from $B_T^{-1} X$ by taking conditional expectations, $E_i = \mathbb{E}_{\mathbb{Q}}(B_T^{-1} X | \mathcal{F}_i)$. By the binomial representation theorem, there is a previsible process ϕ such that

$$E_i = E_0 + \sum_{k=1}^{i} \phi_k \, \Delta Z_k.$$

Now consider the following construction strategy: at tick-time i, buy the portfolio Π_i consisting of:

- ϕ_{i+1} units of the stock S,

- $\psi_{i+1} = (E_i - \phi_{i+1} B_i^{-1} S_i)$ units of the cash bond.

At time zero, our starting point, Π_0 is worth $\phi_1 S_0 + \psi_1 B_0 = E_0 = \mathbb{E}_{\mathbb{Q}}(B_T^{-1} X)$ – it costs that much to create. There is also no difficulty in determining ϕ_1 or ψ_1 as ϕ and ψ are previsible.

What about one tick later? We have held the portfolio safe across the period, but its constituents have changed in value: Π_0 is now worth

$$\phi_1 S_1 + \psi_1 B_1 = B_1\big(E_0 + \phi_1(B_1^{-1} S_1 - B_0^{-1} S_0)\big),$$

but $B_1^{-1} S_1 - B_0^{-1} S_0 = \Delta Z_1$. *Now* we can use the binomial representation theorem to simplify the expression above: at time 1, Π_0 is worth $B_1 E_1$.

We are at time 1, and the construction strategy demands that we buy a new portfolio Π_1. But the portfolio Π_1, which we need to create at time 1, costs precisely that amount above: $B_1 E_1$, *whatever* actually happened to S, that is whichever filtration $\mathcal{F}_1$ actually obtains.

Thus we can cash in our portfolio Π_0 to create Π_1. And so on. At time i, portfolio Π_i costs $B_i E_i$ to purchase, and it will change by time $(i+1)$ to be worth $B_{i+1}E_{i+1}$, the cost of the next portfolio. Our construction strategy is what we might call *self-financing*. And at the end of our self-financing strategy, we end up with the worth of Π_{T-1} at time T, which is $B_T B_T^{-1} X$. That is X, the claim we require.

Arbitrage

The price of the claim X is now obvious: it is $\mathbb{E}_\mathbb{Q}(B_T^{-1} X)$ – the expected value of the discounted claim, under the martingale measure $\mathbb{Q}$ for the discounted stock Z. And it is an arbitrage price because any other price could be milked for free money by running the (ϕ_i, ψ_i) strategy the appropriate way round to duplicate the claim. We shouldn't be too surprised – we are simply repeating the argument of section 2.2 in formal guise. But our formal argument has won us an overview of the entire process and a couple of vital slogans:

The existence of self-financing strategies

The first slogan is that within the binomial tree model, we can produce a self-financing (ϕ_i, ψ_i) strategy which duplicates any claim. What do we mean exactly by self-financing? Let us define V_i, the worth of the trading strategy at time i, to be the opening value of the portfolio Π_i at time i, that is $V_i = \phi_{i+1}S_i + \psi_{i+1}B_i$. Then a strategy is self-financing if the closing value of the portfolio Π_{i-1} at time i is precisely equal to V_i. In symbols, the 'financing gap' of cash that would otherwise have to be injected into the strategy,

$$D_i = V_i - \phi_i S_i - \psi_i B_i,$$

must be zero.

Another way of representing this self-financing property comes from the changes of the strategy value process $\Delta V_i = V_i - V_{i-1}$,

$$\Delta V_i = \phi_i \, \Delta S_i + \psi_i \, \Delta B_i + D_i.$$

The gap D_i at time i is zero if and only if the change in value of the strategy from time $i-1$ to i is due only to *changes in the stock and bond values alone*.

Formally:

Self-financing hedging strategies

Given a binomial tree model of a market with a stock S and bond B, then (ϕ_i, ψ_i) is a self-financing strategy to construct a claim X if:

(i)   both ϕ and ψ are previsible;

(ii)   the change in value V of the portfolio defined by the strategy obeys the difference equation:

$$\Delta V_i = \phi_i \, \Delta S_i + \psi_i \, \Delta B_i$$

where $\Delta S_i := S_i - S_{i-1}$ is the change in S from tick-time $i - 1$ to i, and $\Delta B_i := B_i - B_{i-1}$ is the corresponding change in B;

(iii)   and $\phi_T S_T + \psi_T B_T$ is identically equal to the claim X.

Expectation of the discounted claim under the martingale measure

The second of these slogans is that the price of any derivative within the binomial tree model is the expectation of the discounted claim under the measure $\mathbb{Q}$ which makes the discounted stock a martingale.

Option price formula (discrete case)

The value at tick-time i of a claim X maturing at date T is

$$B_i \, \mathbb{E}_{\mathbb{Q}}\big(B_T^{-1} X \mid \mathcal{F}_i\big).$$

Why? Precisely because there is a self-financing strategy, justified by the binomial representation theorem, which requires that amount to start off and yields the claim without risk at T.

Uniqueness and existence of $\mathbb{Q}$

And in this discrete world, we can add almost as an afterthought that for any sensible stock process S, there will be a unique measure $\mathbb{Q}$ under which $B_i^{-1} S_i$, the discounted stock, is a $\mathbb{Q}$-martingale.

Conclusions

We are now finished in the discrete world, we have the general theorem we require. Any claim on a stock implies a derivative instrument tied to the underlying stock value at any time by a construction strategy capable of providing arbitrage riches if any market player disobeyed it. That arbitrage-justified value is the expectation of the discounted claim, but expectation under just one special measure, the measure $\mathbb{Q}$ under which the discounted stock is a martingale. The real measure $\mathbb{P}$ which S follows is irrelevant. The construction strategy is self-financing and generates the claim whatever S does.

2.4 Overture to continuous models

We can, in a heuristic way, look into the continuous world with our discrete techniques. Without being fully rigorous yet, we could believe that a continuous model can be approximated by a discrete time model with a very small intertick time. Indeed we can show that a natural discrete model with constant growth rate and noise approximates a log–normal distribution under both the original measure $\mathbb{P}$ and the martingale measure $\mathbb{Q}$. It will even be possible to 'derive' the Black–Scholes option pricing formula, though its rigorous development must wait until the very end of the next chapter.

Model with constant stock growth and noise

The model is parameterised by the intertick time δt. As that quantity gets smaller, the model should ever more closely approximate a continuous-time model. There are also three fixed and constant parameters: the noisiness σ, the stock growth rate μ, and the riskless interest rate r.

The cash bond B_t has the simple form that $B_t = \exp(rt)$, which does not depend on the interval size.

The stock process follows the nodes of a recombinant tree, which moves from value s at some particular node along the next up/down branch to the new value

$$\begin{cases} s\exp(\mu\,\delta t + \sigma\sqrt{\delta t}) & \text{if up,} \\ s\exp(\mu\,\delta t - \sigma\sqrt{\delta t}) & \text{if down.} \end{cases}$$

The jumps are all equally likely to be up as down, that is $p = 1/2$ everywhere.

For a fixed time t, if we set n to be the number of ticks till time t, then $n = t/\delta t$ and

$$S_t = S_0 \exp\left(\mu t + \sigma\sqrt{t}\left(\frac{2X_n - n}{\sqrt{n}}\right)\right),$$

where X_n is the total number of the n separate jumps which were up-jumps. The random variable X_n has the binomial distribution with mean $n/2$ and variance $n/4$, so that $(2X_n - n)/\sqrt{n}$ has mean zero and variance one. By the central limit theorem, this distribution converges to that of a normal random variable with zero mean and unit variance. So as δt gets smaller and n gets larger, the distribution of S_t becomes log-normal, as $\log S_t$ is normally distributed with mean $\log S_0 + \mu t$ and variance $\sigma^2 t$.

Under the martingale measure

This is what happens under the original measure $\mathbb{P}$, but what goes on with $\mathbb{Q}$?

Following our formula the martingale measure probability q is

$$q = \frac{s\exp(r\,\delta t) - s_{\text{down}}}{s_{\text{up}} - s_{\text{down}}}.$$

We can calculate that q is approximately equal to

$$q = \tfrac{1}{2}\left(1 - \sqrt{\delta t}\left(\frac{\mu + \tfrac{1}{2}\sigma^2 - r}{\sigma}\right)\right).$$

So, under $\mathbb{Q}$, X_n is still binomially distributed, but now has mean nq and variance $nq(1 - q)$.

Thus $(2X_n - n)/\sqrt{n}$ has mean $-\sqrt{t}(\mu + \tfrac{1}{2}\sigma^2 - r)/\sigma$ and variance asymptotically approaching one. Again the central limit theorem gives the convergence of this to a normal random variable with the same mean and variance exactly one. The corresponding S_t is still log-normally distributed with $\log S_t$ having mean $\log S_0 + (r - \tfrac{1}{2}\sigma^2)t$ and variance $\sigma^2 t$. This can be written

$$S_t = S_0 \exp\left(\sigma\sqrt{t}Z + (r - \tfrac{1}{2}\sigma^2)t\right),$$

where Z is a normal $N(0, 1)$ under $\mathbb{Q}$. We have found the marginal distribution of S_t under the martingale measure $\mathbb{Q}$.

Pricing a call option

If X is the call option maturing at date T, struck at k, with $X = (S_T - k)^+$, then its worth at time zero is

$$\mathbb{E}_{\mathbb{Q}}(B_T^{-1}X) = \mathbb{E}_{\mathbb{Q}}\left[\left(S_0 \exp(\sigma\sqrt{T}Z - \tfrac{1}{2}\sigma^2 T) - k\exp(-rT)\right)^+\right].$$

We will see in chapter three that this evaluates as

$$S_0\Phi\left(\frac{\log\frac{S_0}{k} + (r + \tfrac{1}{2}\sigma^2)T}{\sigma\sqrt{T}}\right) - ke^{-rT}\Phi\left(\frac{\log\frac{S_0}{k} + (r - \tfrac{1}{2}\sigma^2)T}{\sigma\sqrt{T}}\right),$$

where Φ is the normal distribution function $\Phi(x) = \mathbb{Q}(Z \leqslant x)$. This is a preview of the Black–Scholes formula which we shall prove properly in the next chapter.

Chapter 3
Continuous processes

S tock prices are not trees. The discrete trees of the previous chapter are only an approximation to the way that prices actually move. In practice, a price can change at any instant, rather than just at some fixed tick-times when a portfolio can be calmly rebalanced. The binary choice of a single jump 'up' or 'down' only becomes subtle as the ticks get closer and closer, giving the tree more and ever-shorter branches. But such trees grow too complex and we stop being able to see the wood.

We shall have to start from scratch in the continuous world. The discrete models will guide us – the intuitions gained there will be more than useful – but limiting arguments based on letting δt tend to zero are too dangerous to be used rigorously. We will encounter a representation theorem which establishes the basis of risk-free construction and again it will be martingale measures that prime the expectation operator correctly. But processes and measures will be harder to separate intuitively – we will need a calculus to help us. And changes in measure will affect processes in surprising ways. We will no longer be able to proceed in full generality – we will concentrate on Brownian motion and its relatives. If there is one overarching principle to this chapter it is that Brownian motion is sophisticated enough to produce interesting models and simple enough to be tractable. Given the subtleties of working with continuous processes, a simple calculus based on Brownian motion will be more than enough for us.

3.1 Continuous processes

We want randomness. With our discrete stock price model we didn't have any old random process. We forcibly limited ourselves to a binomial tree. We started simply and hoped (with some justification) that complex enough market models could be built from such humble materials. The single binomial branching was the building block for our 'realistic' market. For the continuous world we need an analogous basis – something simple and yet a reasonable starting point for realism.

What is a continuous process? Three small-scale principles guide us. Firstly, the value can change at any time and from moment to moment. Secondly, the actual values taken can be expressed in arbitrarily fine fractions – any real number can be taken as a value. And lastly the process changes continuously – the value cannot make instantaneous jumps. In other words, if the value changes from 1 to 1.05 it must have passed through, albeit quickly, all the values in between.

At least as a starting point, we can insist that stock market indices or prices of individual securities behave this way. Even though they move in a 'sharp-edged' way, it isn't too unrealistic to claim that they nonetheless display continuous process behaviour.

And as far back as Bachelier in 1900, who analysed the motion of the Paris stock exchange, people have gone further and compared the prices to one particular continuous process – the process followed by a randomly moving gas particle, or *Brownian motion* (figure 3.2).

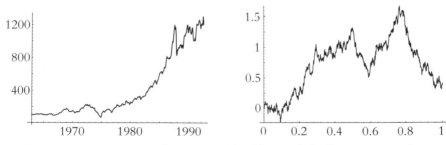

Figure 3.1 UK FTA index, 1963–92 **Figure 3.2** Brownian motion

Locally the likeness can be striking – both display the same jaggedness, and the same similarity under scale changes – the jaggedness never smooths out as the magnification increases. But globally, the similarity fades – figure 3.1

doesn't *look like* figure 3.2. At an intuitive level, the global structure of the stock index is different. It grows, gets 'noisier' as time passes, and doesn't go negative. Brownian motion can't be the whole story.

But we only want a basis – the single binomial branching didn't look promising right away. We shouldn't run ahead of ourselves. Brownian motion will prove a remarkably effective component to build continuous processes with – *locally* Brownian motion looks realistic.

Brownian motion

It was nearly a century after botanist Robert Brown first observed microscopic particles zigzagging under the continuous buffeting of a gas that the mathematical model for their movements was properly developed. The first step to the analysis of Brownian motion is to construct a special family of discrete binomial processes.

The random walk $W_n(t)$

For n a positive integer, define the binomial process $W_n(t)$ to have:

(i) $W_n(0) = 0$,

(ii) layer spacing $1/n$,

(iii) up and down jumps equal and of size $1/\sqrt{n}$,

(iv) measure $\mathbb{P}$, given by up and down probabilities everywhere equal to $\frac{1}{2}$.

In other words, if $X_1, X_2, \ldots$ is a sequence of independent binomial random variables taking values $+1$ or -1 with equal probability, then the value of W_n at the ith step is defined by:

$$W_n\left(\tfrac{i}{n}\right) = W_n\left(\tfrac{i-1}{n}\right) + \frac{X_i}{\sqrt{n}}, \qquad \text{for all } i \geqslant 1.$$

The first two steps are shown in figure 3.3. What does W_n look like as n gets large?

Instead of blowing out of control, the family portraits (figure 3.4) appear to be settling down towards something as n increases. The moves of size

$1/\sqrt{n}$ seem to force some kind of convergence. Can we make a formal statement? Consider for example, the distribution of W_n at time 1: for a particular W_n, there are $n + 1$ possible values that it can take, ranging from $-\sqrt{n}$ to $\sqrt{n}$. But the distribution always has zero mean and unit variance. (Because $W_n(1)$ is the sum of n IID random variables, each with zero mean and variance $1/n$.)

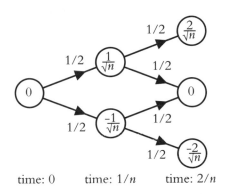

time: 0   time: $1/n$ time: $2/n$

Figure 3.3 The first two steps of the random walk W_n

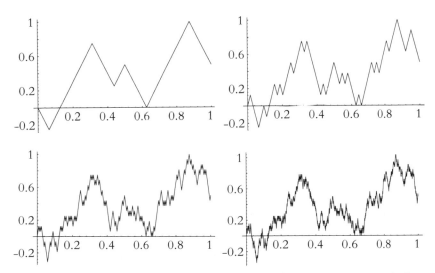

Figure 3.4 Random walks of 16, 64, 256 and 1024 steps respectively

Moreover the central limit theorem gives us a limit for these binomial distributions – as n gets large, the distribution of $W_n(1)$ tends towards the unit normal $N(0, 1)$. In fact, the value of $W_n(t)$ is the same as

$$W_n(t) = \sqrt{t} \left(\frac{\sum_{i=1}^{nt} X_i}{\sqrt{nt}} \right).$$

The distribution of the ratio in brackets tends, by the central limit theorem, to a normal $N(0, 1)$ random variable. And so the distribution of $W_n(t)$ tends to a normal $N(0, t)$.

There is a formal unity underlying the family – all the marginal distributions tend towards the same underlying normal structure.

And not just all the marginal distributions, but all the *conditional* marginal distributions as well. Each random walk W_n has the property that its future movements away from a particular position are independent of where that position is (and indeed independent of its entire history of movements up to that time). Additionally such a future displacement $W_n(s + t) - W_n(s)$ is binomially distributed with zero mean and variance t. Thus again, the central limit theorem gives us a constant limiting structure, and all conditional marginals tend towards a normal distribution of the same mean and variance.

The marginals converge, the conditional marginals converge, and the temptation is irresistible to say that the distributions of the processes converge too. And indeed they do, though this isn't the place to set up the careful formal framework to make sense of that statement. The distribution of W_n converges, and it converges towards *Brownian motion*.

Formally:

Brownian motion

The process $W = (W_t : t \geqslant 0)$ is a $\mathbb{P}$-Brownian motion if and only if

(i) W_t is continuous, and $W_0 = 0$,

(ii)   the value of W_t is distributed, under $\mathbb{P}$, as a normal random variable $N(0, t)$,

(iii) the increment $W_{s+t} - W_s$ is distributed as a normal $N(0, t)$, under $\mathbb{P}$, and is independent of $\mathcal{F}_s$, the history of what the process did up to time s.

These are both necessary and sufficient conditions for the process W to be Brownian motion. The last condition, though an exact echo of the behaviour of the discrete precursors $W_n(t)$, is subtle. Many processes that have marginals $N(0, t)$ are not Brownian motion. In the continuous world, just as it was in the discrete, it is not just the marginals (conditional on the process' value at time zero) that count, but *all* the marginals conditional on *all* the histories $\mathcal{F}_s$. It will in fact be the daunting task of specifying all these that drives us to a Brownian calculus.

Exercise 3.1 If Z is a normal $N(0, 1)$, then the process $X_t = \sqrt{t}Z$ is continuous and is marginally distributed as a normal $N(0, t)$. Is X a Brownian motion?

Exercise 3.2 If W_t and $\tilde{W}_t$ are two independent Brownian motions and ρ is a constant between -1 and 1, then the process $X_t = \rho W_t + \sqrt{1 - \rho^2}\tilde{W}_t$ is continuous and has marginal distributions $N(0, t)$. Is this X a Brownian motion?

It is also worth noting just how *odd* Brownian motion really is. We won't stop to prove them, but here is a brief peek into the bestiary:

- Although W is continuous everywhere, it is (with probability one) differentiable nowhere.

- Brownian motion will eventually hit any and every real value no matter how large, or how negative. It may be a million units above the axis, but it will (with probability one) be back down again to zero, by some later time.

- Once Brownian motion hits a value, it immediately hits it again *infinitely* often, and then again from time to time in the future.

- It doesn't matter what scale you examine Brownian motion on − it looks just the same. Brownian motion is a fractal.

Brownian motion is often also called a Wiener process, and is a (one-dimensional) Gaussian process.

Brownian motion as stock model

We had our misgivings about Brownian motion as a global model for stock behaviour, but we don't have to use it on its own. Brownian motion wanders. It has mean zero, whereas the stock of a company normally grows at some rate – and historically we expect prices to rise if only because of inflation. But we can add in a drift artificially. For example the process $S_t = W_t + \mu t$, for some constant μ reflecting nominal growth, is called Brownian motion with drift.

And if it looks too noisy, or not noisy enough, we can scale the Brownian motion by some factor: for example, $S_t = \sigma W_t + \mu t$, for a constant noise factor σ.

How are we doing? Consider the stock market data shown in figure 3.1. We could estimate σ and μ for the best fit [in this case, $\sigma = 91.3$ and $\mu = 37.8$] and simulate a sample path.

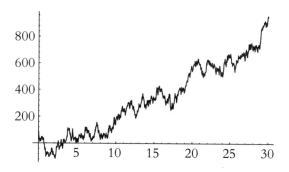

Figure 3.5 Brownian motion plus drift

Not bad – the process has long-term upwards growth, as we want. But in this particular case, we have a glitch right away. The process went negative, which we may not want for the price of a stock of a limited liability company.

Exercise 3.3 Show that, for all values of σ ($\sigma \neq 0$), μ, and $T > 0$ there is always a positive probability that S_T is negative. (Hint: consider the marginal distribution of S_T.)

We can though be more adventurous in shaping Brownian motion to our ends. Consider for example, taking the exponential of our process:

$$X_t = \exp(\sigma W_t + \mu t).$$

Now we mirror the stock market's long-term exponential growth (and for good measure we start off quietly and get noisier). Again finding a best fit for σ and μ [$\sigma = 0.178$ and $\mu = 0.087$, a 'noisiness' of 17.8% and an annual drift of 8.7%] we can simulate a sample path (figure 3.6).

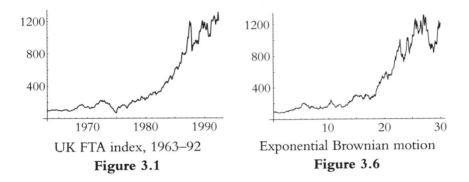

UK FTA index, 1963–92	Exponential Brownian motion
Figure 3.1	**Figure 3.6**

This process is, not surprisingly, well known and it is usually called exponential Brownian motion with drift, or sometimes geometric Brownian motion with drift. It is not the only model for stocks – and indeed we will look at others later on – but it is simple and not that bad. (Could you tell which picture was which without the captions?) Brownian motion can prove an effective building block.

3.2 Stochastic calculus

Shaping Brownian motion with functions may be powerful, but it brings a dangerous complexity. Consider any smooth (differentiable) curve. Globally it can have almost any behaviour it likes, because the condition that it is differentiable does nothing to affect it at a large scale. Suppose we zoom in though, pinning down a small section under a microscope. In figure 3.7, we focus in on the point of a particular differentiable curve with x-co-ordinate of 1.7, increasing the magnification by a factor of about ten each time.

Reading the graphs from left to right and line by line, each small box is expanded to form the frame for the next graph. As the process continues, the graph section becomes smoother and straighter, until eventually it *is* straight – it is a small straight line.

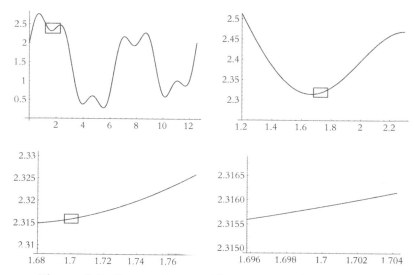

Figure 3.7 Progressive magnification around the point 1.7

Differentiable functions, however strange their global behaviour, are at heart built from straight line segments. Newtonian calculus is the formal acknowledgement of this.

With a Newtonian construction, we could decide to build up a family of nice functions by specifying how they are locally built up out of our building block, the straight line. We would write the change in value of a *Newtonian function f* over a time interval at t of infinitesimal length dt as

$$df_t = \mu_t\, dt,$$

where μ_t is our scaling function, the slope or drift of the magnified straight line at t.

Then we could explore our universe of Newtonian functions. Consider, for example:

(1) The equation $df_t = \mu\, dt$, for some constant μ. What is f? That is, what does it look like? How does it behave globally? Could we draw it? If

we stick together straight line segments of slope μ, then intuitively we just produce a straight line of slope μ. If f_0, for example, was equal to zero then we might guess (correctly) that f_t could be written in more familiar notation as $f_t = \mu t$.

(2)   The equation $df_t = t\,dt$. Here we have a slope at time t of value t — what does this look like? Simple integration comes to the rescue. If $f_0 = 0$, then we could again pin down f_t as $f_t = \frac{1}{2}t^2$. The going was a bit harder here, but we managed it, and we can check it ourselves by differentiation: $f_t' = t$ as we require.

What about uniqueness though? In the first example, our intuition dismissed the possibility of another solution, but what about here? The construction metaphor ($df_t = t\,dt$ tells us how to build f_t, and thus given a starting place and a deterministic building plan we ought to produce just one possible f_t) suggests that $f_t = \frac{1}{2}t^2$ is the unique solution and indeed we can formalise this.

> **Uniqueness of Newtonian differentials**
> Two complementary forms of uniqueness operate here.
>
> - If f_t and $\tilde{f}_t$ are two differentiable functions agreeing at 0 ($f_0 = \tilde{f}_0$) and they have identical drifts ($df_t = d\tilde{f}_t$), then the processes are equal: $f_t = \tilde{f}_t$ for all t. In other words, f is unique given the drift μ_t (and f_0).
>
> - Secondly, given a differentiable function f_t, there is only one drift function μ_t which satisfies $f_t = f_0 + \int_0^t \mu_s\,ds$ (for all t). So μ is unique given f.

Instead of just giving the drift μ_t directly, we might have a problem where the drift itself depends on the current value of the function. Specifically, if the drift μ_t equals $\mu(f_t, t)$, where $\mu(x, t)$ is a known function, then

$$df_t = \mu(f_t, t)\,dt$$

is called an ordinary differential equation (ODE). If there is a differentiable function f which satisfies it (with given f_0), it forms a *solution*. There are plenty of ODEs which have no solutions, and plenty more which do not have unique solutions. (The uniqueness of the solution to an ODE cannot be deduced just from the uniqueness of Newtonian differentials in the box.)

(3)   The equation $df_t = f_t \, dt$. Now things are harder, as direct integration is not a route to the solution. We could guess — say $f_t = e^t$ — and then check by differentiation. This solution happens to be unique for $f_0 = 1$.

(4)   The equation $df_t = f_t t^{-2} \, dt$. This is an example of a bad case, where solutions need neither exist nor be unique. Given $f_0 = 0$, there are an infinite number of solutions, namely $f_t = a \exp(-1/t)$ for every possible value of an arbitrary constant a. However, for $f_0 \neq 0$, there are no solutions at all.

Perhaps our universe of Newtonian functions isn't so benign. It is clear that though ODEs are powerful construction tools, they are also dangerous ones. There are plenty of 'bad' ODEs which we haven't a clue how to explore.

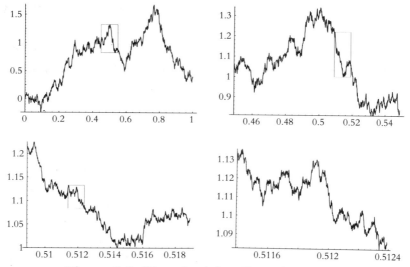

Figure 3.8 'Zooming in' on Brownian motion

Stochastic differentials

And if it was bad for Newtonian differentials, consider a construction procedure based on Brownian motion. Zooming in on Brownian motion doesn't produce a straight line (figure 3.8)

As before, each box is expanded by suitable horizontal and vertical scaling to frame the next graph. The self-similarity of Brownian motion means that each new graph is also a Brownian motion, and just as noisy.

But of course this self-similarity is ideal for a building block – we could build global Brownian motion out of lots of local Brownian motion segments. And we could build general random processes from small segments of Brownian motion (suitably scaled). If we built using straight line segments (suitably scaled) too, we could include Newtonian functions as well.

A stochastic process X will have both a Newtonian term based on dt and a Brownian term, based on the infinitesimal increment of W which we will call dW_t. The Brownian term of X can have a noise factor σ_t, and so the infinitesimal change of X_t is

$$dX_t = \sigma_t \, dW_t + \mu_t \, dt.$$

As in the Newtonian case, the drift μ_t can depend on the time t. But it can also be random and depend on values that X (or indeed W) took up until t itself. And of course, so can the noisiness σ_t. Such processes, like X and σ, whose value at time t can depend on the history $\mathcal{F}_t$, but not the future, are called *adapted* to the filtration $\mathcal{F}$ of the Brownian motion W.

We call σ_t the *volatility* of the process X at time t and μ_t the *drift* of X at t.

Stochastic processes

What does our universe look like? As with Newtonian differentials, finding this out entails 'integrating' stochastic differentials in some way. We can, though, formally define what it is to be a (continuous) stochastic process.

This definition of stochastic process (see box) is not universal, and in particular it excludes discontinuous cases such as Poisson processes. Nevertheless it will be quite adequate for all the models we will meet.

The technical condition that σ and μ must be $\mathcal{F}$-previsible processes means that they are adapted to the filtration $\mathcal{F}$, and that they may have some jump discontinuities. In terms of stochastic analysis, this defines stochastic processes to be semimartingales whose drift term is absolutely continuous. This class is closed under all the operations used later, and all the models considered will lie within it.

And as it happens, we can provide a uniqueness result to mirror the classical setup.

Stochastic processes

A stochastic process X is a continuous process $(X_t : t \geqslant 0)$ such that X_t can be written as

$$X_t = X_0 + \int_0^t \sigma_s \, dW_s + \int_0^t \mu_s \, ds,$$

where σ and μ are random $\mathcal{F}$-previsible processes such that $\int_0^t (\sigma_s^2 + |\mu_s|) \, ds$ is finite for all times t (with probability 1). The differential form of this equation can be written

$$dX_t = \sigma_t \, dW_t + \mu_t \, dt.$$

Uniqueness of volatility and drift

Two complementary forms of uniqueness operate here.

- Firstly, if two processes X_t and $\tilde{X}_t$ agree at time zero ($X_0 = \tilde{X}_0$) and they have identical volatility σ_t and drift μ_t, then the processes are equal: $X_t = \tilde{X}_t$ for all t. In other words, X is unique given σ_t and μ_t (and X_0).

- Secondly, given the process X, there is only one pair of volatility σ_t and drift μ_t which satisfies $X_t = X_0 + \int_0^t \sigma_s \, dW_s + \int_0^t \mu_s \, ds$ (for all t). This uniqueness of σ_t and μ_t given X comes from the *Doob–Meyer decomposition* of semimartingales.

In the special case when σ and μ depend on W only through X_t, such as $\sigma_t = \sigma(X_t, t)$, where $\sigma(x, t)$ is some deterministic function, the equation

$$dX_t = \sigma(X_t, t) \, dW_t + \mu(X_t, t) \, dt$$

is called a stochastic differential equation (SDE) for X. And it will generally be easier to write down the SDE (if it exists) for a particular X then it is to provide an explicit solution for the SDE. As in the Newtonian case (ODEs), an SDE need not have a solution, and if it does it might not be unique. Usage of the term SDE does tend to spread out from this strict definition to include the stochastic differentials of processes whose volatility and drift depends not only on X_t and t, but also on other events in the history $\mathcal{F}_t$.

But can we recognise the world we have created, perhaps in terms of W_t, the Brownian motion we have some handle on?

Partially. In the simple case, where σ and μ are both constants, meaning that X has constant volatility and drift, the SDE for X is

$$dX_t = \sigma \, dW_t + \mu \, dt.$$

It isn't too hard to guess what the solution to this is:

$$X_t = \sigma W_t + \mu t,$$

(assuming that $X_0 = 0$). And our meagre understanding of W_t and dW_t at least gives us some confidence that the differential form of σW_t is $\sigma \, dW_t$. As σ and μ are independent of X, the uniqueness result could form a part of a proof that this is the only solution.

But consider the only slightly more complex SDE (echoing the Newtonian ODE of example (3) above),

$$dX_t = X_t \big(\sigma \, dW_t + \mu \, dt \big).$$

We're at sea.

3.3 Itô calculus

Intuitive integration doesn't carry us very far. We need tools to manipulate the differential equations, just as Newtonian calculus has the chain rule, product rule, integration by parts, and so on.

How far could Newton carry us? Suppose we had some function f of Brownian motion, say $f(W_t) = W_t^2$. Could we use a simple chain rule to produce the stochastic differential df_t? Under Newtonian rules, $d(W_t^2)$ would be $2W_t \, dW_t$, which doesn't look too implausible. But we should check via integration, because

$$\text{if} \quad \int_0^t d(W_s^2) = 2 \int_0^t W_s \, dW_s, \quad \text{then} \quad W_t^2 = 2 \int_0^t W_s \, dW_s.$$

How can we tackle $\int_0^t W_s \, dW_s$? Consider dividing up the time interval $[0, t]$ into a partition $\{0, t/n, 2t/n, \ldots, (n-1)t/n, t\}$ for some n. Then we could approximate the integral with a summation over this partition, that is

$$2 \int_0^t W_s \, dW_s \approx 2 \sum_{i=0}^{n-1} W\left(\tfrac{it}{n}\right) \left(W\left(\tfrac{(i+1)t}{n}\right) - W\left(\tfrac{it}{n}\right) \right).$$

Now something begins to worry us. The difference term inside the brackets is just the increment of Brownian motion from one particular partition point to the next. By property (iii) of Brownian motion, that increment is independent of the Brownian motion up to that point, and in particular it is independent of the Brownian motion term $W(it/n)$. Also the increment has zero mean, which means that so too must the product of the increment and $W(it/n)$. So the summation consists of terms with zero mean, forcing it to have zero mean itself.

But W_t^2 has mean t, because of the variance structure of Brownian motion, so $2W_t \, dW_t$ *cannot* be the differential of W_t^2, because its integral doesn't even have the right expectation.

What went wrong? Consider a Taylor expansion of $f(W_t)$ for some smooth f:

$$df(W_t) = f'(W_t) \, dW_t + \tfrac{1}{2} f''(W_t) \, (dW_t)^2 + \tfrac{1}{3!} f'''(W_t) \, (dW_t)^3 + \ldots$$

Over-familiar with Newtonian differentials, we assumed that $(dW_t)^2$ and higher terms were zero. But as we have observed before, Brownian motion is odd. Take $(dW_t)^2$, given the same partitioning of $[0, t]$ we just used: $\{0, t/n, 2t/n, \ldots, t\}$. We can model the integral of $(dW_t)^2$ by the (hopefully convergent) approximation

$$\int_0^t (dW_t)^2 = \sum_{i=1}^{n} \left(W\left(\tfrac{ti}{n}\right) - W\left(\tfrac{t(i-1)}{n}\right) \right)^2.$$

But if we let $Z_{n,i}$ be

$$Z_{n,i} = \frac{W\left(\tfrac{ti}{n}\right) - W\left(\tfrac{t(i-1)}{n}\right)}{\sqrt{t/n}},$$

then for each n, the sequence $Z_{n,1}, Z_{n,2}, \ldots$ is a set of IID normals $N(0, 1)$. (Because each increment $W\left(\tfrac{ti}{n}\right) - W\left(\tfrac{t(i-1)}{n}\right)$ is a normal $N(0, t/n)$, independent of the ones before it, by Brownian motion fact (iii).)

3.3 Itô calculus

We can rewrite our approximation for $\int (dW_s)^2$ as

$$\int_0^t (dW_s)^2 \approx t \sum_{i=1}^n \frac{Z_{n,i}^2}{n}.$$

By the weak law of large numbers (just like the strong law but only talking about the distribution of random variables), the distribution of the right-hand side summation converges towards the constant expectation of each $Z_{n,i}^2$, namely 1. Thus $\int_0^t (dW_s)^2 = t$, or in differential form $(dW_t)^2 = dt$.

We can't ignore $(dW_t)^2$; it only looks second order because of the notation. What about $(dW_t)^3$ and so on? It turns out that they *are* zero. (For example, $\mathbb{E}(|dW_t|^3)$ has size $(dt)^{3/2}$, which is negligible compared with dt.) So Taylor gives us:

$$df(W_t) = f'(W_t)\,dW_t + \tfrac{1}{2}f''(W_t)\,dt + 0.$$

The formal version of this surprising departure from Newtonian differentials is the deservedly famous *Itô's formula* (sometimes seen modestly as Itô's lemma).

Itô's formula
If X is a stochastic process, satisfying $dX_t = \sigma_t\,dW_t + \mu_t\,dt$, and f is a deterministic twice continuously differentiable function, then $Y_t := f(X_t)$ is also a stochastic process and is given by

$$dY_t = \left(\sigma_t f'(X_t)\right) dW_t + \left(\mu_t f'(X_t) + \tfrac{1}{2}\sigma_t^2 f''(X_t)\right) dt.$$

Returning to our W_t^2, we can apply Itô with $X = W$ and $f(x) = x^2$ and we have

$$d(W_t^2) = 2W_t\,dW_t + dt, \quad \text{or} \quad W_t^2 = 2\int_0^t W_s\,dW_s + t,$$

which at least has the right expectation.

More generally, if X is still just the Brownian motion W, then $f(X)$ has differential

$$df(W_t) = f'(W_t)\,dW_t + \tfrac{1}{2}f''(W_t)\,dt,$$

as hinted above.

 Exercise 3.4 If $X_t = \exp(W_t)$, then what is dX_t?

SDEs from processes

Itô's most immediate use is to generate SDEs from a functional expression for a process. Consider the exponential Brownian motion we set up in section 3.1:

$$X_t = \exp(\sigma W_t + \mu t).$$

What SDE does X follow? We know we can handle the term inside the brackets but we have to take a stochastic differential of the exponential function as well. With the right formulation though, we can use Itô's formula.

Suppose we took Y_t to be the process $\sigma W_t + \mu t$, and f to be the exponential function $f(x) = e^x$. Then Y_t *is* simple enough that we can write down its differential immediately: $dY_t = \sigma\,dW_t + \mu\,dt$. But of course the X_t we want can be written as $X_t = f(Y_t)$, so one application of Itô's formula gives us

$$dX_t = \sigma f'(Y_t)\,dW_t + \left(\mu f'(Y_t) + \tfrac{1}{2}\sigma^2 f''(Y_t)\right) dt.$$

The exponential function is particularly pleasant, as $f'(Y_t) = f''(Y_t) = f(Y_t) = X_t$, so we can rewrite the differential as

$$dX_t = X_t\left(\sigma\,dW_t + (\mu + \tfrac{1}{2}\sigma^2)\,dt\right).$$

Here, the variable σ is sometimes called the *log-volatility* of the process, because it is the volatility of the process $\log X_t$, and which is often abbreviated just to volatility notwithstanding that term's existing definition. We will also use the name *log-drift* for the drift μ of $\log X_t$, which is different from the drift of dX_t/X_t above.

Processes from SDEs

Much like differentiation (easy, but its inverse can be impossible), using Itô to convert processes to SDEs is relatively straightforward. And if that were all we ever wanted to do there would be few problems. But it isn't – one of the key needs we have is to go in the opposite direction and convert SDEs to processes. Or in other words, to solve them.

In general we can't. Most stochastic differential equations are just too difficult to solve. But a few, rare examples can be, and just like some ODEs they depend on an inspired guess and then a proof that the proposed solution is an actual solution via Itô. Such a solution to an SDE is called a *diffusion*.

Suppose we are asked to solve the SDE

$$dX_t = \sigma X_t \, dW_t.$$

We need an inspired guess – so we notice that the stochastic term ($\sigma X_t \, dW_t$) from this SDE is the same as the SDE we generated via Itô in the section above. Moreover, *if* we choose μ to be $-\frac{1}{2}\sigma^2$, then the drift term in the SDE would match our SDE as well. We guess then that

$$X_t = \exp(\sigma W_t - \tfrac{1}{2}\sigma^2 t).$$

What does Itô tell us? That dX_t is indeed $\sigma X_t \, dW_t$, which is what we wanted. So we have found *one* solution, and as it turns out, the only solution (up to constant multiples). Soluble SDEs are scarce, and this one is special enough to have a name: the Doléans exponential of Brownian motion.

Let's go back then to the SDE we tripped over earlier:

$$dX_t = X_t \big(\sigma \, dW_t + \mu \, dt \big).$$

We could match both drift and volatility terms for this SDE and the SDE of $\exp(\sigma W_t + \nu t)$ if and only if we take ν to be $\mu - \frac{1}{2}\sigma^2$. So that is our guess, that

$$X_t = \exp\big(\sigma W_t + (\mu - \tfrac{1}{2}\sigma^2)t\big).$$

And again Itô confirms our intuition.

 Exercise 3.5 What is the solution of $dX_t = X_t(\sigma \, dW_t + \mu_t \, dt)$, for μ_t a general bounded integrable function of time?

The product rule

Another Newtonian law was the product rule, that $d(f_t g_t) = f_t \, dg_t + g_t \, df_t$. In the stochastic world, there are two (seemingly) separate cases.

In the more significant case, X_t and Y_t are adapted to the same Brownian motion W, in that

$$dX_t = \sigma_t \, dW_t + \mu_t \, dt,$$
$$dY_t = \rho_t \, dW_t + \nu_t \, dt.$$

By applying Itô's formula to $\frac{1}{2}\big((X_t + Y_t)^2 - X_t^2 - Y_t^2\big) = X_t Y_t$, we can see that

$$d(X_t Y_t) = X_t \, dY_t + Y_t \, dX_t + \sigma_t \rho_t \, dt.$$

The final term above is actually $dX_t \, dY_t$ (following from $(dW_t)^2 = dt$) and again marks the difference between Newtonian and Itô calculus.

In the other case, X_t and Y_t are two stochastic processes adapted to two different and independent Brownian motions, such as

$$dX_t = \sigma_t \, dW_t + \mu_t \, dt,$$
$$dY_t = \rho_t \, d\tilde{W}_t + \nu_t \, dt,$$

where σ_t and ρ_t are the respective volatilities of X and Y, μ_t and ν_t are their drifts, and W and $\tilde{W}$ are two independent Brownian motions. Here

$$d(X_t Y_t) = X_t \, dY_t + Y_t \, dX_t,$$

just as in the Newtonian case.

At a deeper level these two stochastic cases can be reconciled by viewing X_t and Y_t as both adapted to the two-dimensional Brownian motion $(W_t, \tilde{W}_t)$, as will be explained in section 6.3.

 Exercise 3.6 Show that if B_t is a zero-volatility process and X_t is any stochastic process, then

$$d(B_t X_t) = B_t \, dX_t + X_t \, dB_t.$$

3.4 Change of measure – the C-M-G theorem

Something remains hidden from us. One of the central themes of the previous chapter was the importance of separating process and measure. Yet we don't seem to mention measures in our stochastic differentials. We may have our basic tools for manipulating stochastic processes, but they are a manipulation of differentials of Brownian motion, not a manipulation of measure. We haven't actually ignored the importance of measure – W_t is not strictly a Brownian motion *per se*, but a Brownian motion with respect to some measure $\mathbb{P}$, a $\mathbb{P}$–Brownian motion. And thus our stochastic differential formulation describes the behaviour of the process X with respect to the measure $\mathbb{P}$ that makes the W_t (or of course the dW_t) a Brownian motion. But the only tool we have seen so far gives us no clue how W_t let alone X_t changes as the measure changes.

As it happens, Brownian motions change in easy and pleasant ways under changes in measure. And thus by extension through their differentials, so do stochastic processes.

Change of measure – the Radon–Nikodym derivative

To get some intuitive feel for the effects of a change of measure, we should go back for a while to discrete processes. Consider a simple two-step random walk:

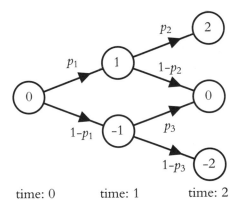

time: 0   time: 1   time: 2

Figure 3.9 Two-step recombinant tree

To get from time 0 to time 2, we can follow four possible paths $\{0, 1, 2\}$,

$\{0, 1, 0\}$, $\{0, -1, 0\}$ and $\{0, -1, -2\}$. Suppose we specified the probability of taking these paths:

Table 3.1 Path probabilities

Path	Probability	
$\{0, 1, 2\}$	$p_1 p_2$	$=: \pi_1$
$\{0, 1, 0\}$	$p_1 (1 - p_2)$	$=: \pi_2$
$\{0, -1, 0\}$	$(1 - p_1) p_3$	$=: \pi_3$
$\{0, -1, -2\}$	$(1 - p_1)(1 - p_3)$	$=: \pi_4$

We could view this mapping of paths to path probabilities as encoding the measure $\mathbb{P}$. If we knew π_1, π_2, π_3 and π_4, then (as long as all of them are strictly between 0 and 1) we know p_1, p_2 and p_3. Thus if we represent our process with a non-recombining tree, we can label each of the paths at the end with the π-information encoding the measure.

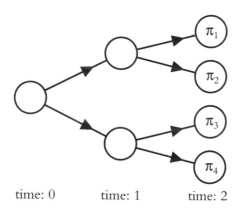

time: 0   time: 1   time: 2

Figure 3.10 Tree with path probabilities marked

Now suppose we had a different measure $\mathbb{Q}$ with probabilities q_1, q_2 and q_3. Again we can code this up with path probabilities, say π_1', π_2', π_3' and π_4'. And again if each π' is strictly between 0 and 1, π_1', π_2', π_3' and π_4' uniquely decides $\mathbb{Q}$.

And with this encoding, there is a very natural way of encoding the differences between $\mathbb{P}$ and $\mathbb{Q}$, giving some idea of how to distort $\mathbb{P}$ so as to produce $\mathbb{Q}$. If we form the ratio π_i'/π_i for each path i, we write the mapping of paths to this ratio as $\frac{dQ}{dP}$. This random variable (random because it depends on the path) is called the *Radon–Nikodym derivative* of $\mathbb{Q}$ with respect to $\mathbb{P}$ up to time 2.

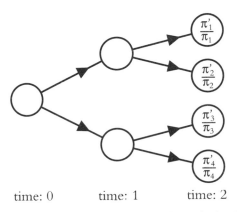

time: 0   time: 1   time: 2

Figure 3.11 Tree with Radon–Nikodym derivative marked

From $\frac{dQ}{dP}$ we can derive $\mathbb{Q}$ from $\mathbb{P}$. How? If we have $\mathbb{P}$, then we have $\pi_1, \pi_2, \ldots, \pi_4$, and $\frac{dQ}{dP}$ gives us the ratios π_i'/π_i, so we have $\pi_1', \pi_2', \ldots, \pi_4'$. And thus $\mathbb{Q}$.

What about p_i or q_i being zero or one? Two things happen – firstly it can become impossible to back out the p_i from the π_i. Consider if p_1 is zero then both π_1 and π_2 are zero and so information about p_2 is lost. But then of course, the paths corresponding to π_1 and π_2 are both impossible (probability zero), so in some sense p_2 really isn't relevant. If we restrict ourselves to only providing π_i for possible paths, then we *can* recover the corresponding p's.

The second problem has a similar flavour but is more serious. Suppose one of the p's is zero, but none of the q's are. Then at least one π_i will be zero when none of the π_i' are. Not all the ratios π_i'/π_i will be well defined, and thus $\frac{dQ}{dP}$ can't exist. We could suppress those paths which had path probability zero, but now we *have* lost something. Those paths may have been $\mathbb{P}$-impossible but they are $\mathbb{Q}$-possible. If we throw them away, then we have lost information about $\mathbb{Q}$ just where it is relevant – paths which are $\mathbb{Q}$-possible. Somehow we can't define $\frac{dQ}{dP}$ if $\mathbb{Q}$ allows something which $\mathbb{P}$ doesn't. And of course *vice versa*.

This is important enough to formalise.

Equivalence

Two measures $\mathbb{P}$ and $\mathbb{Q}$ are *equivalent* if they operate on the same sample space and agree on what is possible. Formally, if A is any event in the sample space,

$$\mathbb{P}(A) > 0 \iff \mathbb{Q}(A) > 0.$$

In other words, if A is possible under $\mathbb{P}$ then it is possible under $\mathbb{Q}$, and if A is impossible under $\mathbb{P}$ then it is also impossible under $\mathbb{Q}$. And *vice versa*.

We can only meaningfully define $\frac{d\mathbb{Q}}{d\mathbb{P}}$ and $\frac{d\mathbb{P}}{d\mathbb{Q}}$ if $\mathbb{P}$ and $\mathbb{Q}$ are equivalent, and then only where paths are $\mathbb{P}$-possible. But of course if paths are $\mathbb{P}$-impossible then we know how $\mathbb{Q}$ acts on those paths – if $\mathbb{Q}$ is equivalent to $\mathbb{P}$ then they are $\mathbb{Q}$-impossible as well.

Thus two measures $\mathbb{P}$ and $\mathbb{Q}$ must be equivalent before they will have Radon–Nikodym derivatives $\frac{d\mathbb{Q}}{d\mathbb{P}}$ and $\frac{d\mathbb{P}}{d\mathbb{Q}}$.

Expectation and $\frac{d\mathbb{Q}}{d\mathbb{P}}$

While we are still working with discrete processes, we should stock up on some facts about expectation and the Radon–Nikodym derivative. One of the reasons for defining it was the efficient coding it represented. Everything we needed to know about $\mathbb{Q}$ could be extracted from $\mathbb{P}$ and $\frac{d\mathbb{Q}}{d\mathbb{P}}$.

Consider then a claim X known by time 2 on our discrete two-step process. The claim X is a random variable, or in other words a mapping from paths to values – we can let x_i denote the value the claim takes if path i is followed. So the expectation of X with respect to $\mathbb{P}$ is given by

$$\mathbb{E}_{\mathbb{P}}(X) = \sum_i \pi_i x_i,$$

where i ranges over all four possible paths. And the expectation of X with respect to $\mathbb{Q}$ is

$$\mathbb{E}_{\mathbb{Q}}(X) = \sum_i \pi_i' x_i = \sum_i \pi_i \left(\frac{\pi_i'}{\pi_i} x_i \right) = \mathbb{E}_{\mathbb{P}} \left(\frac{d\mathbb{Q}}{d\mathbb{P}} X \right).$$

Just like X, $\frac{d\mathbb{Q}}{d\mathbb{P}}$ is a random variable which we can take the expectation of. And the conversion from $\mathbb{Q}$ to $\mathbb{P}$ is pleasingly simple: $\mathbb{E}_{\mathbb{Q}}(X) = \mathbb{E}_{\mathbb{P}}(\frac{d\mathbb{Q}}{d\mathbb{P}} X)$.

Attractive though this is, it represents just one simple case: $\frac{dQ}{dP}$ is defined with a particular time horizon in mind – the ends of the paths, in this case $T = 2$. We specified X at this time and we only wanted an unconditioned expectation. In formal terms, the result we derived was

$$\mathbb{E}_Q\left(X_T \mid \mathcal{F}_0\right) = \mathbb{E}_P\left(\frac{dQ}{dP}X_T \mid \mathcal{F}_0\right),$$

where T is the time horizon for $\frac{dQ}{dP}$ and X_T is known at time T. What about $\mathbb{E}_Q(X_t|\mathcal{F}_s)$ for t not equal to T and s not equal to zero? We need somehow to know $\frac{dQ}{dP}$ not just for the ends of paths but everywhere – $\frac{dQ}{dP}$ is a random variable, but we would like a process.

Radon–Nikodym process

We can do this by letting the time horizon vary, and setting ζ_t to be the Radon–Nikodym derivative taken up to the horizon t. That is, ζ_t is the Radon–Nikodym derivative $\frac{dQ}{dP}$ but only following paths up to time t, and only looking at the ratio of probabilities up to that time. For instance, at time 1, the possible paths are $\{0, 1\}$ and $\{0, -1\}$ and the derivative ζ_1 has values on them of q_1/p_1 and $(1 - q_1)/(1 - p_1)$ respectively. At time zero, the derivative process ζ_0 is just 1, as the only 'path' is the point $\{0\}$ which has probability 1 under both $\mathbb{P}$ and $\mathbb{Q}$. Concretely, we can fill in ζ_t on our tree in terms of the p's and q's (figure 3.12).

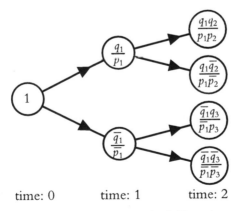

time: 0　　　　　time: 1　　　　　time: 2

Figure 3.12 Tree with ζ_t process marked ($\bar{p}_i = 1 - p_i$, $\bar{q}_i = 1 - q_i$)

In fact there is another expression for ζ_t as the conditional expectation of the T-horizon Radon–Nikodym derivative,

$$\zeta_t = \mathbb{E}_\mathbb{P}\left(\frac{dQ}{d\mathbb{P}}\,\middle|\,\mathcal{F}_t\right),$$

for every t less than or equal to the horizon T.

 Exercise 3.7 Prove that this equation holds for $t = 0, 1, 2$.

We can see that the expectation with respect to $\mathbb{P}$ unpicks the $\frac{dQ}{d\mathbb{P}}$ in just the right way. The process ζ_t represents just what we wanted – an idea of the amount of change of measure so far up to time t along the current path. If we wanted to know $\mathbb{E}_Q(X_t)$ it would be $\mathbb{E}_\mathbb{P}(\zeta_t X_t)$, where X_t is a claim known at time t. If we want to know $\mathbb{E}_Q(X_t|\mathcal{F}_s)$ then we need the amount of change of measure from time s to time t – which is just ζ_t/ζ_s. That is, the change up to time t with the change up to time s removed. In other words

$$\mathbb{E}_Q\left(X_t\,\middle|\,\mathcal{F}_s\right) = \zeta_s^{-1}\mathbb{E}_\mathbb{P}\left(\zeta_t X_t\,\middle|\,\mathcal{F}_s\right).$$

 Exercise 3.8 Prove this on the tree.

Radon-Nikodym summary

Given $\mathbb{P}$ and Q equivalent measures and a time horizon T, we can define a random variable $\frac{dQ}{d\mathbb{P}}$ defined on $\mathbb{P}$-possible paths, taking positive real values, such that

(i) $\mathbb{E}_Q(X_T) = \mathbb{E}_\mathbb{P}\left(\dfrac{dQ}{d\mathbb{P}}X_T\right),$ for all claims X_T knowable by time T.

(ii) $\mathbb{E}_Q(X_t\,|\,\mathcal{F}_s) = \zeta_s^{-1}\mathbb{E}_\mathbb{P}\left(\zeta_t X_t\,|\,\mathcal{F}_s\right),$ $s \leqslant t \leqslant T,$

where ζ_t is the process $\mathbb{E}_\mathbb{P}(\frac{dQ}{d\mathbb{P}}|\mathcal{F}_t)$.

Change of measure – the continuous Radon–Nikodym derivative

What now? To define a measure for Brownian motion it seems we have to be able to write down the likelihood of every possible path the process can take, ranging across not only a continuous-valued state space but also a continuous-valued time line. Standard probability theory gives some clue to the technology required, if we were content merely to represent the marginal distributions for the process at each time. Despite the continuous nature of the state space, we know that we can express likelihoods in terms of a probability density function.

For example, the measure $\mathbb{P}$ on the real numbers, corresponding to a normal $N(0,1)$ random variable X, can be represented via the density $f_{\mathbb{P}}(x)$, where

$$f_{\mathbb{P}}(x) = \frac{1}{\sqrt{2\pi}} e^{-\frac{1}{2}x^2}.$$

In some loose sense, $f_{\mathbb{P}}(x)$ represents the relative likelihood of the event $\{X = x\}$ occurring. Or, less informally the probability that X lies between x and $x + dx$ is approximately $f_{\mathbb{P}}(x)dx$. In exact terms, the probability that X takes a value in some subset A of the reals is

$$\mathbb{P}(X \in A) = \int_A \frac{1}{\sqrt{2\pi}} e^{-\frac{1}{2}x^2} \, dx.$$

For example, the chance of X being in the interval $[0, 1]$ is the integral of the density over the interval, $\int_0^1 f_{\mathbb{P}}(x) \, dx$, which has value 0.3413.

But marginal distributions aren't enough – a single marginal distribution won't capture the nature of the process (we can see that clearly even on a discrete tree). Nor will all the marginal distributions for each time t. We need nothing less than all the marginal distributions at each time t *conditional on every history* $\mathcal{F}_s$ for all times $s < t$. We need to capture the idea of a likelihood of a path in the continuous case, by means of some conceptual handle on a particular path specified for all times $t < T$.

One approach is to specify a path if not for all times before the horizon T, then at least for some arbitrarily large yet still finite set of times $\{t_0 = 0, t_1, \ldots, t_{n-1}, t_n = T\}$. Consider then, the set of paths which go through the points $\{x_1, \ldots, x_n\}$ at times $\{t_1, \ldots, t_n\}$. If there were just one time t_1 and one point x_1, then we could write down the likelihood of such a path. We could use the probability density function of W_{t_1}, $f_{\mathbb{P}}^1(x)$, which is the density function of a normal $N(0, t_1)$, or

$$f_{\mathbb{P}}^1(x) = \frac{1}{\sqrt{2\pi t_1}} \exp\left(-\frac{x^2}{2t_1}\right).$$

And if we can do this for one time t_1, then we can for finitely many t_i. All we require is the joint likelihood function $f_{\mathbb{P}}^n(x_1, \ldots, x_n)$ for the process taking values $\{x_1, \ldots, x_n\}$ at times $\{t_1, \ldots, t_n\}$.

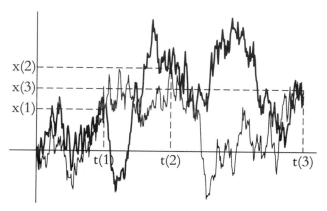

Figure 3.13 Two Brownian motions agreeing on the set $\{t_1, t_2, t_3\}$

Joint likelihood function for Brownian motion

If we take t_0 and x_0 to be zero, and write Δx_i for $x_i - x_{i-1}$ and $\Delta t_i = t_i - t_{i-1}$, then given the third condition of Brownian motion that increments $\Delta W_i = W(t_i) - W(t_{i-1})$ are mutually independent, we can write down

$$f_{\mathbb{P}}^n(x_1, \ldots, x_n) = \prod_{i=1}^n \frac{1}{\sqrt{2\pi\Delta t_i}} \exp\left(-\frac{(\Delta x_i)^2}{2\Delta t_i}\right).$$

So we can write down a likelihood function corresponding to the measure $\mathbb{P}$ for a process on a finite set of times. And in the continuous limit, we have a handle on the measure $\mathbb{P}$ for a continuous process. If A is some subset of $\mathbb{R}^n$, then the $\mathbb{P}$-probability that the random n-vector $(W_{t_1}, \ldots, W_{t_n})$ is in A is exactly the integral over A of the likelihood function $f_{\mathbb{P}}^n$.

Radon–Nikodym derivative – continuous version
Suppose $\mathbb{P}$ and $\mathbb{Q}$ are equivalent measures. Given a path ω, for every ordered time mesh $\{t_1, \ldots, t_n\}$ (with $t_n - T$), we define x_i to be $W_{t_i}(\omega)$, and then the derivative $\frac{d\mathbb{Q}}{d\mathbb{P}}$ up to time T is defined to be the limit of the likelihood ratios

$$\frac{d\mathbb{Q}}{d\mathbb{P}}(\omega) = \lim_{n \to \infty} \frac{f_{\mathbb{Q}}^n(x_1, \ldots, x_n)}{f_{\mathbb{P}}^n(x_1, \ldots, x_n)},$$

as the mesh becomes dense in the interval $[0, T]$.

This continuous-time derivative $\frac{d\mathbb{Q}}{d\mathbb{P}}$ still satisfies the results that

(i)
$$\mathbb{E}_{\mathbb{Q}}(X_T) = \mathbb{E}_{\mathbb{P}}\left(\frac{d\mathbb{Q}}{d\mathbb{P}} X_T\right),$$

(ii)
$$\mathbb{E}_{\mathbb{Q}}(X_t \mid \mathcal{F}_s) = \zeta_s^{-1} \mathbb{E}_{\mathbb{P}}(\zeta_t X_t \mid \mathcal{F}_s), \qquad s \leqslant t \leqslant T,$$

where ζ_t is the process $\mathbb{E}_{\mathbb{P}}(\frac{d\mathbb{Q}}{d\mathbb{P}} | \mathcal{F}_t)$, and X_t is any process adapted to the history $\mathcal{F}_t$.

Just as the measure $\mathbb{P}$ can be approached through a limiting time mesh, so can the Radon–Nikodym derivative $\frac{d\mathbb{Q}}{d\mathbb{P}}$. The event of paths agreeing with ω on the mesh, $A = \{\omega' : W_{t_i}(\omega') = W_{t_i}(\omega), \ i = 1, \ldots, n\}$, gets smaller and smaller till it is just the single point-set $\{\omega\}$. The Radon–Nikodym derivative can be thought of as the limit

$$\frac{d\mathbb{Q}}{d\mathbb{P}}(\omega) = \lim_{A \to \{\omega\}} \frac{\mathbb{Q}(A)}{\mathbb{P}(A)}.$$

Simple changes of measure – Brownian motion plus constant drift
We have the mechanics of change of measure but still no clue about what change of measure does in the continuous world. Suppose, for example, we had a $\mathbb{P}$-Brownian motion W_t. What does W_t *look* like under an equivalent measure $\mathbb{Q}$ – is it still recognisably Brownian motion or something quite different?

Foresight can provide one simple example. Consider W_t a $\mathbb{P}$-Brownian motion, then (out of nowhere) define $\mathbb{Q}$ to be a measure equivalent to $\mathbb{P}$ via

$$\frac{d\mathbb{Q}}{d\mathbb{P}} = \exp\left(-\gamma W_T - \tfrac{1}{2}\gamma^2 T\right),$$

for some time horizon T. What does W_t look like with respect to $\mathbb{Q}$?

One place to start, and it is just a start, is to look at the marginal of W_T under $\mathbb{Q}$. We need to find the likelihood function of W_T with respect to $\mathbb{Q}$, or something equivalent. One useful trick is to look at moment-generating functions:

> **Identifying normals**
>
> A random variable X is a normal $N(\mu, \sigma^2)$ under a measure $\mathbb{P}$ if and only if
>
> $$\mathbb{E}_{\mathbb{P}}\big(\exp(\theta X)\big) = \exp\big(\theta\mu + \tfrac{1}{2}\theta^2\sigma^2\big), \qquad \text{for all real } \theta.$$

To calculate $\mathbb{E}_{\mathbb{Q}}\big(\exp(\theta W_T)\big)$, we can use fact (i) of the Radon–Nikodym derivative summary, which tells us that it is the same as the $\mathbb{P}$-expectation $\mathbb{E}_{\mathbb{P}}\big(\frac{d\mathbb{Q}}{d\mathbb{P}}\exp(\theta W_T)\big)$. This equals

$$\mathbb{E}_{\mathbb{P}}\big(\exp(-\gamma W_T - \tfrac{1}{2}\gamma^2 T + \theta W_T)\big) = \exp\big(-\tfrac{1}{2}\gamma^2 T + \tfrac{1}{2}(\theta - \gamma)^2 T\big),$$

because W_T is a normal $N(0, T)$ with respect to $\mathbb{P}$.

Simplifying the algebra, we have

$$\mathbb{E}_{\mathbb{Q}}\big(\exp(\theta W_T)\big) = \exp(-\theta\gamma T + \tfrac{1}{2}\theta^2 T),$$

which is the moment-generating function of a normal $N(-\gamma T, T)$. Thus the marginal distribution of W_T, under $\mathbb{Q}$, is also a normal with variance T but with mean $-\gamma T$.

What about W_t for t less than T? The marginal distribution of W_T is what we would expect if W_t under $\mathbb{Q}$ were a Brownian motion plus a constant drift $-\gamma$. Of course, a lot of other process also have a marginal normal $N(-\gamma T, T)$ distribution at time T, but it would be an elegant result if the sole effect of changing from $\mathbb{P}$ to $\mathbb{Q}$ via $\frac{d\mathbb{Q}}{d\mathbb{P}} = \exp(-\gamma W_T - \tfrac{1}{2}\gamma^2 T)$ were just to punch in a drift of $-\gamma$.

And so it is. The process W_t is a Brownian motion with respect to $\mathbb{P}$ and Brownian motion with constant drift $-\gamma$ under $\mathbb{Q}$. Using our two results about $\frac{d\mathbb{Q}}{d\mathbb{P}}$, we can prove the three conditions for $\tilde{W}_t = W_t + \gamma t$ to be $\mathbb{Q}$-Brownian motion:

(i) $\tilde{W}_t$ is continuous and $\tilde{W}_0 = 0$;

(ii) $\tilde{W}_t$ is a normal $N(0,t)$ under $\mathbb{Q}$;

(iii) $\tilde{W}_{t+s} - \tilde{W}_s$ is a normal $N(0,t)$ independent of $\mathcal{F}_s$.

The first of these is true and (ii) and (iii) can be re-expressed as

(ii)′ $\mathbb{E}_{\mathbb{Q}}\big(\exp(\theta \tilde{W}_t)\big) = \exp(\tfrac{1}{2}\theta^2 t)$;

(iii)′ $\mathbb{E}_{\mathbb{Q}}\big(\exp\big(\theta(\tilde{W}_{t+s} - \tilde{W}_s)\big) \mid \mathcal{F}_s\big) = \exp(\tfrac{1}{2}\theta^2 t)$.

 Exercise 3.9 Show that (ii)′ and (iii)′ are equivalent to (ii) and (iii) respectively, and prove them using the change of measure process $\zeta_t = \mathbb{E}_{\mathbb{P}}(\frac{d\mathbb{Q}}{d\mathbb{P}}|\mathcal{F}_t)$.

That both W_t and $\tilde{W}_t$ are Brownian motion, albeit with respect to different measures, seems paradoxical. But switching from $\mathbb{P}$ to $\mathbb{Q}$ just changes the relative likelihood of a particular path being chosen. For example, W *might* follow a path which drifts downwards for a time at a rate of about $-\gamma$. Although that path is $\mathbb{P}$-unlikely, it is $\mathbb{P}$-possible. Under $\mathbb{Q}$, on the other hand, such a path is much more likely, and the chances are that is what we see. But it still could be just improbable Brownian motion behaviour.

We can see this in the Radon–Nikodym derivative $\frac{d\mathbb{Q}}{d\mathbb{P}}$ which is large when W_T is very negative, and small when W_T is closer to zero or positive. This is just the consequence of the common sense thought that paths which end up negative are more likely under $\mathbb{Q}$ (Brownian motion plus downward drift) than they are under $\mathbb{P}$ (driftless Brownian motion). Correspondingly, paths which finish near or above zero are less likely under $\mathbb{Q}$ than $\mathbb{P}$.

Cameron–Martin–Girsanov

So this one change of measure just changed a vanilla Brownian motion into one with drift – nothing else. And of course, drift is one of the elements of our stochastic differential form of processes. In fact *all* that measure changes on Brownian motion can do is to change the drift. All the processes that we are interested in are representable as instantaneous differentials made up of some amount of Brownian motion and some amount of drift. The mapping of stochastic differentials under $\mathbb{P}$ to stochastic differentials under $\mathbb{Q}$ is both natural and pleasing.

This is what our theorem provides.

Cameron–Martin–Girsanov theorem

If W_t is a $\mathbb{P}$-Brownian motion and γ_t is an $\mathcal{F}$-previsible process satisfying the boundedness condition $\mathbb{E}_{\mathbb{P}} \exp\left(\frac{1}{2} \int_0^T \gamma_t^2 \, dt\right) < \infty$, then there exists a measure $\mathbb{Q}$ such that

(i) $\mathbb{Q}$ is equivalent to $\mathbb{P}$

(ii) $\dfrac{d\mathbb{Q}}{d\mathbb{P}} = \exp\left(-\int_0^T \gamma_t \, dW_t - \frac{1}{2} \int_0^T \gamma_t^2 \, dt\right)$

(iii) $\tilde{W}_t = W_t + \int_0^t \gamma_s \, ds$ is a $\mathbb{Q}$-Brownian motion.

In other words, W_t is a drifting $\mathbb{Q}$-Brownian motion with drift $-\gamma_t$ at time t.

Within constraints, if we want to turn a $\mathbb{P}$-Brownian motion W_t into a Brownian motion with some specified drift $-\gamma_t$, then there's a $\mathbb{Q}$ which does it.

Within limits, drift is measure and measure drift.

Conversely to the theorem,

Cameron–Martin–Girsanov converse

If W_t is a $\mathbb{P}$-Brownian motion, and $\mathbb{Q}$ is a measure equivalent to $\mathbb{P}$, then there exists an $\mathcal{F}$-previsible process γ_t such that

$$\tilde{W}_t = W_t + \int_0^t \gamma_s \, ds$$

is a $\mathbb{Q}$-Brownian motion. That is, W_t plus drift γ_t is $\mathbb{Q}$-Brownian motion. Additionally the Radon–Nikodym derivative of $\mathbb{Q}$ with respect to $\mathbb{P}$ (at time T) is $\exp(-\int_0^T \gamma_t \, dW_t - \frac{1}{2} \int_0^T \gamma_t^2 \, dt)$.

C-M-G and stochastic differentials

The C-M-G theorem applies to Brownian motion, but all our processes are disguised Brownian motions at heart. Now we can see the rewards of our Brownian calculus instantly – C-M-G becomes a powerful tool for controlling the drift of *any* process.

3.4 Change of measure – the C-M-G theorem

Suppose that X is a stochastic process with increment

$$dX_t = \sigma_t \, dW_t + \mu_t \, dt,$$

where W is a $\mathbb{P}$-Brownian motion. Suppose we want to find if there is a measure $\mathbb{Q}$ such that the drift of process X under $\mathbb{Q}$ is $\nu_t \, dt$ instead of $\mu_t \, dt$. As a first step, dX can be rewritten as

$$dX_t = \sigma_t \left(dW_t + \left(\frac{\mu_t - \nu_t}{\sigma_t} \right) dt \right) + \nu_t \, dt.$$

If we set γ_t to be $(\mu_t - \nu_t)/\sigma_t$, and if γ then satisfies the C-M-G growth condition ($\mathbb{E}_\mathbb{P} \exp(\frac{1}{2} \int_0^T \gamma_t^2 \, dt) < \infty$) then indeed there is a new measure $\mathbb{Q}$ such that $\tilde{W}_t := W_t + \int_0^t (\mu_s - \nu_s)/\sigma_s \, ds$ is a $\mathbb{Q}$-Brownian motion.

But this means that the differential of X under $\mathbb{Q}$ is

$$dX_t = \sigma_t \, d\tilde{W}_t + \nu_t \, dt,$$

where $\tilde{W}$ is a $\mathbb{Q}$-Brownian motion – which gives X the drift ν_t we wanted.

We can also set limits on the changes that changing to an equivalent measure can wreak on a process. Since the change of measure can only change the Brownian motion to a Brownian motion plus drift, the volatility of the process must remain the same.

Examples – changes of measure

1. Let X_t be the drifting Brownian process $\sigma W_t + \mu t$, where W is a $\mathbb{P}$-Brownian motion and σ and μ are both constant. Then using C-M-G with $\gamma_t = \mu/\sigma$, there exists an equivalent measure $\mathbb{Q}$ under which $\tilde{W}_t = W_t + (\mu/\sigma)t$ and $\tilde{W}$ is a $\mathbb{Q}$-Brownian motion up to time T. Then $X_t = \sigma \tilde{W}_t$, which is (scaled) $\mathbb{Q}$-Brownian motion.
   The measures also give rise to different expectations. For example, $\mathbb{E}_\mathbb{P}(X_t^2)$ equals $\mu^2 t^2 + \sigma^2 t$, but $\mathbb{E}_\mathbb{Q}(X_t^2) = \sigma^2 t$.

2. Let X_t be the exponential Brownian motion with SDE

$$dX_t = X_t(\sigma \, dW_t + \mu \, dt),$$

where W is $\mathbb{P}$-Brownian motion. Can we change measure so that X has the new SDE

$$dX_t = X_t(\sigma \, dW_t + \nu \, dt),$$

for some arbitrary constant drift ν?

Using C–M–G with $\gamma_t = (\mu - \nu)/\sigma$, there is indeed a measure $\mathbb{Q}$ under which $\tilde{W}_t = W_t + (\mu - \nu)t/\sigma$ is a $\mathbb{Q}$-Brownian motion. Then X does have the SDE

$$dX_t = X_t(\sigma \, d\tilde{W}_t + \nu \, dt),$$

where $\tilde{W}$ is a $\mathbb{Q}$-Brownian motion.

3.5 Martingale representation theorem

We can solve some SDEs with Itô; we can see how SDEs change as measure changes. But central to answering our pricing question in chapter two was the concept of a measure with respect to which the process was expected to stay the same, the *martingale measure* for our discrete trees. The price of derivatives turned out to be an expectation under this measure, and the construction of this expectation even showed us the trading strategy required to justify this price. And so it is here.

First the description again:

Martingales

A stochastic process M_t is a *martingale* with respect to a measure $\mathbb{P}$ if and only if

(i) $\mathbb{E}_\mathbb{P}\big(|M_t|\big) < \infty, \qquad$ for all t

(ii) $\mathbb{E}_\mathbb{P}\big(M_t \mid \mathcal{F}_s\big) = M_s, \qquad$ for all $s \leqslant t$.

The first condition is merely a technical sweetener, it is the second that carries the weight. A martingale measure is one which makes the expected future value conditional on its present value and past history merely its present value. It isn't *expected* to drift upwards or downwards.

Some examples:

(1)   Trivially, the constant process $S_t = c$ (for all t) is a martingale with respect to any measure: $\mathbb{E}_\mathbb{P}(S_t|\mathcal{F}_s) = c = S_s$, for all $s \leqslant t$, and for any

measure $\mathbb{P}$.

(2)   Less trivially, $\mathbb{P}$-Brownian motion is a $\mathbb{P}$-martingale. Intuitively this makes sense – Brownian motion doesn't move consistently up or down, it's as likely to do either. But we should get into the habit of checking this formally: we need $\mathbb{E}_{\mathbb{P}}(W_t|\mathcal{F}_s) = W_s$. Of course we have that the increment $W_t - W_s$ is independent of $\mathcal{F}_s$ and distributed as a normal $N(0, t-s)$, so that $\mathbb{E}_{\mathbb{P}}(W_t - W_s|\mathcal{F}_s) = 0$. This yields the result, as

$$\mathbb{E}_{\mathbb{P}}(W_t|\mathcal{F}_s) = \mathbb{E}_{\mathbb{P}}(W_s|\mathcal{F}_s) + \mathbb{E}_{\mathbb{P}}(W_t - W_s|\mathcal{F}_s) = W_s + 0.$$

(3)   For any claim X depending only on events up to time T, the process $N_t = \mathbb{E}_{\mathbb{P}}(X|\mathcal{F}_t)$ is a $\mathbb{P}$-martingale (assuming only the technical constraint $\mathbb{E}_{\mathbb{P}}(|X|) < \infty$).

Example (3) is an elegant little trick for producing martingales – and as we shall see (and have already seen in chapter two) central to pricing derivatives. First why? Convince yourself that $N_t = \mathbb{E}_{\mathbb{P}}(X|\mathcal{F}_t)$ is a well-defined *process* – the first stage of the alchemy is the introduction of a time line into the random variable X. Now for N_t to be a $\mathbb{P}$-martingale, we require $\mathbb{E}_{\mathbb{P}}(N_t|\mathcal{F}_s) = N_s$, but for this we merely need to be satisfied that

$$\mathbb{E}_{\mathbb{P}}\left(\mathbb{E}_{\mathbb{P}}(X|\mathcal{F}_t)\,\Big|\,\mathcal{F}_s\right) = \mathbb{E}_{\mathbb{P}}\left(X\,|\,\mathcal{F}_s\right).$$

That is, that conditioning firstly on information up to time t and then on information up to time s is just the same as conditioning up to time s to begin with. This property of conditional expectation is the *tower law*.

Exercise 3.10 Show that the process $X_t = W_t + \gamma t$, where W_t is a $\mathbb{P}$-Brownian motion, is a $\mathbb{P}$-martingale if and only if $\gamma = 0$.

Representation

In chapter two, we had a binomial representation theorem – if M_t and N_t are both $\mathbb{P}$-martingales then they share more than just the name – locally they can only differ by a scaling, by the size of the opening of each particular branching. We could represent *changes* in N_t by scaled *changes* in the other non-trivial $\mathbb{P}$-martingale. Thus N_t itself can be represented by the scaled *sum* of these changes.

In the continuous world:

> **Martingale representation theorem**
> Suppose that M_t is a $\mathbb{Q}$-martingale process, whose volatility σ_t satisfies the additional condition that it is (with probability one) always non-zero. Then if N_t is any other $\mathbb{Q}$-martingale, there exists an $\mathcal{F}$-previsible process ϕ such that $\int_0^T \phi_t^2 \sigma_t^2 \, dt < \infty$ with probability one, and N can be written as
>
> $$N_t = N_0 + \int_0^t \phi_s \, dM_s.$$
>
> Further ϕ is (essentially) unique.

This is virtually identical to the earlier result, with summation replaced by an integral. As we are getting used to, the move to a continuous process extracts a formal technical penalty. In this case, the $\mathbb{Q}$-martingale's volatility must be positive with probability 1 – but otherwise our chapter two result has carried across unchanged. If there is a measure $\mathbb{Q}$ under which M_t is a $\mathbb{Q}$-martingale, then any other $\mathbb{Q}$-martingale can be represented in terms of M_t. The process ϕ_t is simply the ratio of their respective volatilities.

Driftlessness

We need just one more tool. Thrown into the discussion of martingales was the intuitive description of a martingale as neither drifting up or drifting down. We have, though, a technical definition of drift via our stochastic differential formulation. An obvious question springs to mind: are stochastic processes with no drift term always martingales, and *vice versa* can martingales always be represented as just $\sigma_t \, dW_t$ for some $\mathcal{F}$-previsible volatility process σ_t?

Nearly.

One way round we can do for ourselves with the martingale representation theorem. If a process X_t is a $\mathbb{P}$-martingale then with W_t a $\mathbb{P}$-Brownian motion, we have an $\mathcal{F}$-previsible process ϕ_t such that

$$X_t = X_0 + \int_0^t \phi_s \, dW_s.$$

This is just the integral form of the increment $dX_t = \phi_t\, dW_t$, which has no drift term.

The other way round is true (up to a technical constraint), but harder. For reference:

> **A collector's guide to martingales**
> If X is a stochastic process with volatility σ_t (that is $dX_t = \sigma_t\, dW_t + \mu_t\, dt$) which satisfies the technical condition $\mathbb{E}\big[(\int_0^T \sigma_s^2\, ds)^{\frac{1}{2}}\big] < \infty$, then
>
> $$X \text{ is a martingale} \iff X \text{ is driftless } (\mu_t \equiv 0).$$

If the technical condition fails, a driftless process may not be a martingale. Such processes are called *local martingales*.

Exponential martingales

The technical constraint can be tiresome. For example, take the (driftless) SDE for an exponential process $dX_t = \sigma_t X_t\, dW_t$. The condition (in this case, $\mathbb{E}\big[(\int_0^T \sigma_s^2 X_s^2\, ds)^{\frac{1}{2}}\big] < \infty$) is difficult to check, but for these specific exponential examples, a better (more practical) test is:

> **A collector's guide to exponential martingales**
> If $dX_t = \sigma_t X_t\, dW_t$, for some $\mathcal{F}$-previsible process σ_t, then
>
> $$\mathbb{E}\Big(\exp\big(\tfrac{1}{2}\int_0^T \sigma_s^2\, ds\big)\Big) < \infty \;\Rightarrow\; X \text{ is a martingale.}$$

We also note that the solution to the SDE is $X_t = X_0 \exp\big(\int_0^t \sigma_s\, dW_s - \tfrac{1}{2}\int_0^t \sigma_s^2\, ds\big)$.

 Exercise 3.11 If σ_t is a bounded function of both time and sample path, show that $dX_t = \sigma_t X_t\, dW_t$ is a $\mathbb{P}$-martingale.

3.6 Construction strategies

We have the mathematical tools – Itô, Cameron–Martin–Girsanov, and the martingale representation theorem – now we need some idea of how to hook them into a financial model. In the simplest models, Black–Scholes for example, we'll have a market consisting of one random security and a riskless cash account bond; and with this comes the idea of a portfolio.

The portfolio (ϕ, ψ)

A *portfolio* is a pair of processes ϕ_t and ψ_t which describe respectively the number of units of security and of the bond which we hold at time t. The processes can take positive or negative values (we'll allow unlimited short-selling of the stock or bond). The security component of the portfolio ϕ should be $\mathcal{F}$-*previsible*: depending only on information up to time t but not t itself.

There is an intuitive way to think about previsibility. If ϕ were left-continuous (that is, ϕ_s tends to ϕ_t as s tends upwards to t from below) then ϕ would be previsible. If ϕ were only right-continuous (that is, ϕ_s tends to ϕ_t only as s tends downwards to t from above), then ϕ need not be.

Self-financing strategies

With the idea of a portfolio comes the idea of a strategy. The description (ϕ_t, ψ_t) is a dynamic strategy detailing the amount of each component to be held at each instant. And one particularly interesting set of strategies or portfolios are those that are financially self-contained or *self financing*.

A portfolio is self-financing if and only if the change in its value only depends on the change of the asset prices. In the discrete framework this was captured via a difference equation, and in the continuous case it is equivalent to an SDE.

What SDE?

With stock price S_t and bond price B_t, the value, V_t, of a portfolio (ϕ_t, ψ_t) at time t is given by $V_t = \phi_t S_t + \psi_t B_t$. At the next time instant, two things happen: the old portfolio changes value because S_t and B_t have changed price; and the old portfolio has to be adjusted to give a new portfolio as instructed by the trading strategy (ϕ, ψ). If the cost of the adjustment is perfectly matched by the profits or losses made by the portfolio then no extra money is required from outside – the portfolio is self-financing.

In our discrete language, we had the difference equation

$$\Delta V_i = \phi_i \, \Delta S_i + \psi_i \, \Delta B_i.$$

In continuous time, we get a stochastic differential equation:

Self-financing property

If (ϕ_t, ψ_t) is a portfolio with stock price S_t and bond price B_t, then

$$(\phi_t, \psi_t) \quad \text{is self-financing} \iff dV_t = \phi_t \, dS_t + \psi_t \, dB_t.$$

Suppose the stock price S_t is given by a simple Brownian motion W_t (so $S_t = W_t$ for all t), and the bond price B_t is constant ($B_t = 1$ for all t). What kind of portfolios are self-financing?

(1) Suppose $\phi_t = \psi_t = 1$ for all t. If we hold a unit of stock and a unit of bond for all time without change, then the value of the portfolio ($V_t = W_t + 1$) may fluctuate, but it will all be due to fluctuation of the stock. Intuitively, no extra money is needed to come in to uphold the (ϕ_t, ψ_t) strategy and none comes out – this (ϕ_t, ψ_t) portfolio ought to be self-financing.

Checking this formally, $V_t = W_t + 1$ implies that $dV_t = dW_t$ which is the same as $\phi_t \, dS_t + \psi_t \, dB_t$, as we required (remembering that $dB_t = 0$).

(2) Suppose $\phi_t = 2W_t$ and $\psi_t = -t - W_t^2$. Here (ϕ_t, ψ_t) is a portfolio, ϕ_t is previsible, and the value $V_t = \phi_t S_t + \psi_t B_t = W_t^2 - t$. By Itô's formula, $dV_t = 2W_t \, dW_t$ which is identical to $\phi_t \, dS_t + \psi_t \, dB_t$ as required.

 Exercise 3.12 Verify the Itô claim in (2) above (which also shows that $W_t^2 - t$ is a martingale).

Surprising though it seems: holding as many units of stock as twice its current price, though a rollercoaster strategy, is exactly offset by the stock profits and the changing bond holding of $-(t + W_t^2)$. The (ϕ_t, ψ_t) strategy could (in a perfect market) be followed to our heart's content without further funding.

The second example should convince us that being self-financing is not an automatic property of a portfolio. The Itô check worked, but it could easily have failed if ψ_t had been different – the (ϕ_t, ψ_t) strategy would have required injections or forced outflows of cash. Every time we claim a portfolio is self-financing we have to turn the handle on Itô's formula to check the SDE.

Trading strategies

Now we can define a replicating strategy for a claim:

Replicating strategy

Suppose we are in a market of a riskless bond B and a risky security S with volatility σ_t, and a claim X on events up to time T.

A *replicating strategy* for X is a self-financing portfolio (ϕ, ψ) such that $\int_0^T \sigma_t^2 \phi_t^2 \, dt < \infty$ and $V_T = \phi_T S_T + \psi_T B_T = X$.

Why should we care about replicating strategies? For the same reason as we wanted them in the discrete market models. The claim X gives the value of some derivative which we need to pay off at time T. We want a price if there is one, as of now, given a model for S and B.

If there is a replicating strategy (ϕ_t, ψ_t), then the price of X at time t *must be* $V_t = \phi_t S_t + \psi_t B_t$. (And specifically, the price at time zero is $V_0 = \phi_0 S_0 + \psi_0 B_0$.) If it were lower, a market player could buy one unit of the derivative at time t and sell ϕ_t units of S and ψ_t units of B against it, continuing to be short (ϕ, ψ) until time T. Because (ϕ, ψ) is self-financing and the portfolio is worth X at time T guaranteed, the bought derivative and sold portfolio would safely cancel at time T, and no extra money is required between times t and T. The profit created by the mismatch at time t can be banked there and then without risk. And, as usual with arbitrage, one unit could have been many; no risk means no fear.

And of course if the derivative price had been higher than V_t, then we could have *sold* the derivative and *bought* the self-financing (ϕ, ψ) to the same effect. Replicating strategies, *if* they exist, tie down the price of the claim X not just at payoff but everywhere.

We can lay out a battle plan. We define a market model with a stock price process complex enough to satisfy our need for realism. Then, using whatever tools we have to hand we find replicating strategies for all useful claims X. And if we can, we can price derivatives in the model. The rest of the book consists of upping the stakes in complexity of models and of claims.

3.7 Black–Scholes model

We need a model to cut our teeth on. We have the tools and we've seen the overall approach at the end of chapter two. So taking the stock model of section 3.1, we will use the Cameron–Martin–Girsanov theorem (section 3.4) to change it into a martingale, and then use the martingale representation theorem (section 3.5) to create a replicating strategy for each claim. Itô will oil the works.

The model

Our first model – basic Black–Scholes

We will posit the existence of a deterministic r, μ and σ such that the bond price B_t and the stock price follow

$$B_t = \exp(rt),$$
$$S_t = S_0 \exp(\sigma W_t + \mu t),$$

where r is the *riskless interest rate*, σ is the *stock volatility* and μ is the *stock drift*. There are no transaction costs and both instruments are freely and instantaneously tradable either long or short at the price quoted.

We need a model for the behaviour of the stock – simple enough that we actually can find replicating strategies but not so simple that we can't bring ourselves to believe in it as a model of the real world.

Following in Black and Scholes' footsteps, our market will consist of a riskless constant-interest rate cash bond and a risky tradable stock following an exponential Brownian motion.

As we've seen in section 3.1, it is at least a plausible match to the real world. And as we shall see here, it is quite hard enough to start with.

Zero interest rates

If there's one parameter that throws up a smokescreen around a first run at an analysis of the Black–Scholes model, it's the interest rate r. The problems it causes are more tedious than fatal – as we'll see soon, the tools we have are powerful enough to cope. But we'll temporarily simplify things, and set r to be zero.

So now we begin. For an arbitrary claim X, knowable by some horizon time T, we want to see if we can find a replicating strategy (ϕ_t, ψ_t).

Finding a replicating strategy

We shall follow a three-step process outlined in this box here.

Three steps to replication

(1) Find a measure $\mathbb{Q}$ under which S_t is a martingale.

(2) Form the process $E_t = \mathbb{E}_{\mathbb{Q}}(X|\mathcal{F}_t)$.

(3) Find a previsible process ϕ_t, such that $dE_t = \phi_t \, dS_t$.

The tools described earlier on will be essential to do this. We shall use the Cameron–Martin–Girsanov theorem (section 3.4) for the first step and the martingale representation theorem (section 3.5) for the third one.

Step one

For two different reasons – firstly we need to apply the Cameron–Martin–Girsanov theorem, and secondly we need to be able to tell if S_t is a $\mathbb{Q}$-martingale for a given $\mathbb{Q}$ – we want to find an SDE for S_t.

The stock follows an exponential Brownian motion, $S_t = \exp(\sigma W_t + \mu t)$, so the logarithm of the stock price, $Y_t = \log(S_t)$, follows a simple drifting Brownian motion $Y_t = \sigma W_t + \mu t$. Thus the SDE for Y_t is easy to write down: $dY_t = \sigma\, dW_t + \mu\, dt$. But, of course, Itô makes it possible to write down the SDE for $S_t = \exp(Y_t)$ as

$$dS_t = \sigma S_t\, dW_t + (\mu + \tfrac{1}{2}\sigma^2)S_t\, dt.$$

In order for S_t to be a martingale, the first thing to do is to kill the drift in this SDE. If we let γ_t be a process with constant value $\gamma = (\mu + \tfrac{1}{2}\sigma^2)/\sigma$, then the C-M-G theorem says that there is a measure $\mathbb{Q}$ such that $\tilde{W}_t = W_t + \gamma t$ is $\mathbb{Q}$-Brownian motion. (The technical boundedness condition is satisfied because γ_t is constant.) Substituting in, the SDE is now

$$dS_t = \sigma S_t\, d\tilde{W}_t.$$

No drift term, thus S_t could be a $\mathbb{Q}$-martingale. The exponential martingales box (section 3.5) contains a condition in terms of σ for S_t to be a martingale under $\mathbb{Q}$. As σ is constant, the condition holds which means that S_t must be a $\mathbb{Q}$-martingale. Consequently, $\mathbb{Q}$ is the *martingale measure* for S_t.

Step two

Given $\mathbb{Q}$, we can convert X into a process by forming $E_t = \mathbb{E}_{\mathbb{Q}}(X|\mathcal{F}_t)$. This is, as we have already discussed in example (3) of section 3.5, a $\mathbb{Q}$-martingale.

Step three

Since there is a $\mathbb{Q}$, under which both E_t and S_t are $\mathbb{Q}$-martingales, we can invoke the martingale representation theorem. There exists a previsible process ϕ_t which constructs $E_t = \mathbb{E}_{\mathbb{Q}}(X|\mathcal{F}_t)$ out of S_t. (To use the theorem, we need to check that the volatility of S_t is always positive, but this is true because the volatility is just σS_t, and both σ and S_t are always positive.) Formally:

$$E_t = \mathbb{E}_{\mathbb{Q}}(X|\mathcal{F}_t) = \mathbb{E}_{\mathbb{Q}}(X) + \int_0^t \phi_s\, dS_s,$$

or, of course, $dE_t = \phi_t\, dS_t$. So the martingale representation theorem tells us an important fact: given a $\mathbb{Q}$ that makes S_t a $\mathbb{Q}$-martingale with positive volatility, $dE_t = \phi_t\, dS_t$ for some ϕ_t.

We need a replicating strategy (ϕ_t, ψ_t), and it's tempting to believe that we have got one half of it. So we should try it, setting ψ_t to be the only thing it can be, given that we want the portfolio to be worth E_t for all t.

Replicating strategy

Our strategy is to:

- hold ϕ_t units of stock at time t and

- hold $\psi_t = E_t - \phi_t S_t$ units of the bond at time t.

Is it self-financing? The value of the portfolio at time t is

$$V_t = \phi_t S_t + \psi_t B_t = E_t,$$

because the bond B_t is constantly equal to 1. Thus $dV_t = dE_t$, but of course $dE_t = \phi_t\, dS_t$, from the martingale representation theorem.

Since dB_t is zero, we have the self-financing condition we want, namely $dV_t = \phi_t\, dS_t + \psi_t\, dB_t$.

Since the terminal value of the strategy V_T is $E_T = X$, we have a replicating strategy for X – which means there is an arbitrage price for X at all times. Specifically there is an arbitrage price for X at time zero – the value of the (ϕ_t, ψ_t) portfolio at time zero, which makes the price E_0, or $\mathbb{E}_\mathbb{Q}(X)$. In other words, the price of the claim X is its expected value under the measure that makes the stock process S_t a martingale.

It is worth pausing to let a few surprises sink in. The first is just the fact that there are replicating strategies for arbitrary claims. The model that we have chosen isn't too unrealistic – it has the right kind of behaviour and a healthy degree of randomness. So we might expect to fail in our search for replicating strategies. It is after all particularly odd that despite the lack of knowledge about the claim's eventual value, we can nevertheless trade in the market in such a way that we always produce it.

The second surprise, and just as important, is that the price of the derivative has such a simple expression – the expected value of the claim. It is the easiest thing to forget that this is *not* the expectation of the claim with respect to the real measure of S_t, which is the measure that makes it an exponential Brownian motion with drift μ and volatility σ. All *that* expectation could give us would be a long-term average of the claim's payout. And though that could be a useful thing to know in order to judge whether punting with the derivative is worthwhile in the long run, it doesn't give a price. There is a replicating strategy and thus an arbitrage price for the claim. And arbitrage always wins out.

The price happens to be an expectation, but not *the* expectation in a traditional statistics sense. It could only be *the* expectation if quite by chance the drift μ we believe in for the stock were exactly and precisely right to make S_t a martingale in the first place ($\mu = -\frac{1}{2}\sigma^2$).

The third surprise is the simplicity of the process S_t under its martingale measure. If we actually want to crank the handle and calculate derivative prices for a particular claim, we have to be able to calculate the expected value of the claim under the martingale measure $\mathbb{Q}$. Since the claim depends on S_t, this normally involves calculating the expected value, under $\mathbb{Q}$, of some function of the values of S_t up to $t = T$. If S_t were an unpleasant process under $\mathbb{Q}$, then this task could be unpleasant too. But S_t is *also* an exponential Brownian motion under $\mathbb{Q}$. If we solve the SDE, then

$$S_t = \exp(\sigma \tilde{W}_t - \tfrac{1}{2}\sigma^2 t),$$

and we find that S_t has the same constant volatility σ and a new but also constant drift of $-\frac{1}{2}\sigma^2$. So if we felt that S_t was tractable under its original measure, it is also tractable under the martingale measure.

Non-zero interest rates

Now we can bring the interest rate r back in again. What happens if r is non-zero? We can't just ignore it. Suppose we did, and considered a forward contract with claim $S_T - k$ for some price k. We already know that the k which gives the forward contract a zero value at time zero is $k = S_0 e^{rT}$. The arbitrage to produce this is easy to figure out. But our rule, when r was zero, of simply taking the expected value of the claim under the martingale measure for S_t cannot work. In fact,

$$\mathbb{E}_\mathbb{Q}\big(S_T - S_0 e^{rT}\big) = S_0\big(1 - e^{rT}\big) \neq 0.$$

Even discounting the claim won't help in this case. So our rule of finding a measure which makes S_t into a martingale only holds true when r is zero. When r is not zero, the inexorable growth of cash gets in the way.

So we take a guess. If the growth of cash is annoying, simply remove it by discounting everything. We call B_t^{-1} the discount process, and form a discounted stock $Z_t = B_t^{-1} S_t$ and a discounted claim $B_T^{-1} X$.

In this discounted world, we could be forgiven for thinking that r was zero again. So maybe our analysis will work again. Of course, this is all just heuristic justification, and the proof is only in the doing. If we can't find a replicating strategy then, attractive as our guess is, it is also wrong.

Fortunately, we can. Focusing on our discounted stock process Z_t, it is not too hard to write down an SDE

$$dZ_t = Z_t\left(\sigma\, dW_t + (\mu - r + \tfrac{1}{2}\sigma^2)\, dt\right).$$

 Exercise 3.13 Prove it.

Step one

To make Z_t into a martingale, we can invoke C-M-G just as before, only now to introduce a drift of $(\mu - r + \tfrac{1}{2}\sigma^2)/\sigma$ to the underlying Brownian motion. So there exists (another) $\mathbb{Q}$ equivalent to the original measure $\mathbb{P}$ and a $\mathbb{Q}$–Brownian motion $\tilde{W}_t$ such that

$$dZ_t = \sigma Z_t\, d\tilde{W}_t.$$

So Z_t, under $\mathbb{Q}$, is driftless and a martingale.

Step two

We need a process which hits the discounted claim and is also a $\mathbb{Q}$-martingale. And, as before, conditional expectation provides it, namely by forming the process $E_t = \mathbb{E}_{\mathbb{Q}}\left(B_T^{-1} X \mid \mathcal{F}_t\right)$.

Step three

The discounted stock price Z_t is a $\mathbb{Q}$-martingale; and so is the conditional expectation process of the discounted claim E_t. Thus the martingale representation theorem gives us a previsible ϕ_t such that $dE_t = \phi_t\, dZ_t$.

We want to hit the real claim with amounts of the real stock, but in our shadow discounted world we can hit the discounted claim by holding ϕ_t units of the discounted stock. So just as a guess, let us try ϕ_t out in the real world as well.

What about the bond holding? The bond holding in the discounted world is $\psi_t = E_t - \phi_t Z_t$, so we can try that in the real world too. Some reassurance comes from the fact that at time T we will be holding ϕ_T units of the stock and ψ_T units of the bond which will be worth $\phi_T S_T + \psi_T B_T = B_T E_T = X$.

So our replicating strategy is to

- hold ϕ_t units of the stock at time t, and

- hold $\psi_t = E_t - \phi_t Z_t$ units of the bond.

Are we right? The value V_t of the portfolio (ϕ_t, ψ_t) is given by $V_t = \phi_t S_t + \psi_t B_t = B_t E_t$. Thus following exercise 3.6, we can write dV_t as

$$dV_t = B_t \, dE_t + E_t \, dB_t.$$

But dE_t is $\phi_t \, dZ_t$ (our fact from the martingale representation theorem), and so $dV_t = \phi_t B_t \, dZ_t + E_t \, dB_t$. A bit of rearrangement tells us that $E_t = \phi_t Z_t + \psi_t$, and thus

$$dV_t = \phi_t B_t \, dZ_t + (\phi_t Z_t + \psi_t) \, dB_t = \phi_t (B_t \, dZ_t + Z_t \, dB_t) + \psi_t \, dB_t.$$

But, from exercise 3.6 again, $d(B_t Z_t) = B_t \, dZ_t + Z_t \, dB_t$, and since $S_t = B_t Z_t$, we have

$$dV_t = \phi_t \, dS_t + \psi_t \, dB_t.$$

That is, (ϕ_t, ψ_t) is self-financing.

Self-financing strategies

A portfolio strategy (ϕ_t, ψ_t) of holdings in a stock S_t and a non-volatile cash bond B_t has value $V_t = \phi_t S_t + \psi_t B_t$ and discounted value $E_t = \phi_t Z_t + \psi_t$, where Z is the discounted stock process $Z_t = B_t^{-1} S_t$. Then the strategy is self-financing if either

$$dV_t = \phi_t \, dS_t + \psi_t \, dB_t,$$

or equivalently $\quad dE_t = \phi_t \, dZ_t.$

A strategy is self-financing if changes in its value are due only to changes in the assets' values, or equivalently if changes in its discounted value are due only to changes in the discounted values of the assets.

Since we know that $V_T = X$, then we have proved that (ϕ_t, ψ_t) is a replicating strategy for X. Our guesses came good.

Summary

Suppose we have a Black–Scholes model for a continuously tradable stock and bond, that is assuming the existence of a constant r, μ and σ such that their respective prices can be represented as $S_t = S_0 \exp(\sigma W_t + \mu t)$ and $B_t = \exp(rt)$. Then all integrable claims X, knowable by some time horizon T, have associated replicating strategies (ϕ_t, ψ_t). In addition, the arbitrage price of such a claim X is given by

$$V_t = B_t \, \mathbb{E}_\mathbb{Q}\big(B_T^{-1} X \mid \mathcal{F}_t\big) = e^{-r(T-t)} \mathbb{E}_\mathbb{Q}(X \mid \mathcal{F}_t),$$

where $\mathbb{Q}$ is the martingale measure for the discounted stock $B_t^{-1} S_t$.

The important measure $\mathbb{Q}$ is not the measure which makes the stock a martingale, but the measure that makes the *discounted* stock a martingale. And the arbitrage price of the claim is the expectation under $\mathbb{Q}$ of the *discounted* claim.

So when interest rates are non-zero, what are the new rules? They are just discounted versions of the old rules:

Three steps to replication (discounted case)

(1)   Find a measure $\mathbb{Q}$ under which the discounted stock price Z_t is a martingale.

(2)   Form the process $E_t = \mathbb{E}_\mathbb{Q}\big(B_T^{-1} X \mid \mathcal{F}_t\big)$.

(3)   Find a previsible process ϕ_t, such that $dE_t = \phi_t \, dZ_t$.

Call options

We should price something. Following Black and Scholes, we'll price a call option – the right but not the obligation to buy a unit of stock for a predetermined amount at a particular exercise date, say T. If we let this predetermined amount be k (in financial terms, the *strike* of the option), then in formal notation, our claim is $\max(S_T - k, 0)$. Or in more convenient notation, $(S_T - k)^+$.

First we should find V_0, the value of the replicating strategy (and thus the option) at time zero. Our formula tells us that this is given by

$$e^{-rT} \mathbb{E}_{\mathbb{Q}}\left((S_T - k)^+\right),$$

where $\mathbb{Q}$ is the martingale measure for $B_t^{-1} S_t$.

But how do we find this? The first thing to notice is the simplicity of the claim. The value $(S_T - k)^+$ only depends on the stock price at one point in time – namely the expiry time, T. So to find the expectation of this claim we need only find the marginal distribution of S_T under $\mathbb{Q}$.

And to do that, we can look at the process for S_t written in terms of the $\mathbb{Q}$-Brownian motion $\tilde{W}_t$. Since $d(\log S_t) = \sigma \, d\tilde{W}_t + (r - \frac{1}{2}\sigma^2) \, dt$, if we denote the stock price at time zero, S_0, by s, we have that $\log S_t = \log s + \sigma \tilde{W}_t + (r - \frac{1}{2}\sigma^2)t$, and thus $S_t = s \exp\left(\sigma \tilde{W}_t + (r - \frac{1}{2}\sigma^2)t\right)$.

So the marginal distribution for S_T is given by s times the exponential of a normal with mean $(r - \frac{1}{2}\sigma^2)T$ and variance $\sigma^2 T$. Thus if we let Z be a normal $N(-\frac{1}{2}\sigma^2 T, \sigma^2 T)$, we can write S_T as $se^{(Z+rT)}$ and thus the claim as the expectation $e^{-rT}\mathbb{E}\left((se^{(Z+rT)} - k)^+\right)$, which equals

$$\frac{1}{\sqrt{2\pi\sigma^2 T}} \int_{\log(k/s) - rT}^{\infty} (se^x - ke^{-rT}) \exp\left(-\frac{(x + \frac{1}{2}\sigma^2 T)^2}{2\sigma^2 T}\right) dx.$$

This integral can be decomposed by a change of variables into a couple of standard cumulative normal integrals. If we use the notation $\Phi(x)$ to denote $(2\pi)^{-\frac{1}{2}} \int_{-\infty}^{x} \exp(-y^2/2) \, dy$, the probability that a normal $N(0,1)$ has value less than x, then we can calculate that $V_0 = V(s, T)$, where

Black–Scholes formula

$$V(s, T) = s\Phi\left(\frac{\log \frac{s}{k} + (r + \frac{1}{2}\sigma^2)T}{\sigma\sqrt{T}}\right) - ke^{-rT}\Phi\left(\frac{\log \frac{s}{k} + (r - \frac{1}{2}\sigma^2)T}{\sigma\sqrt{T}}\right).$$

This is the Black–Scholes formula for pricing European call options. (Put options, the right to sell a unit of stock for k, can be priced as a call less a forward – put-call parity.)

Exercise 3.14 Find the change of variable and thus prove the Black–Scholes formula.

3.8 Black–Scholes in action

If a stock has a constant volatility of 18% and constant drift of 8%, with continuously compounded interest rates constant at 6%, what is the value of an option to buy the stock for $25 in two years time, given a current stock price of $20?

The description fits the Black–Scholes conditions. Thus using $s = 20$, $k = 25$, $\sigma = 0.18$, $r = 0.06$, and $t = 2$, we can calculate V_0 as $1.221.

 Exercise 3.15 What information about the drift was required?

Price dependence

For values of the current stock price s much smaller than the exercise price k, the value of the formula itself gets small, signifying that the option is out of the money and unlikely to recover in time. Conversely, for values of s much greater than k, the option loses most of its optionality, and becomes a forward. Correspondingly the option price is approximately $s - ke^{-rT}$, which is the current value of a stock forward struck at price k for time T.

Time dependence

As the time to maturity T gets smaller, the chances of the price moving much more decreases and the option value gets closer and closer to the claim value taken at the current price, $(s - k)^+$.

For larger times, however, the option value gets larger. An option with almost infinite time to maturity would have value approaching s, as the cost now of price k is almost zero. It can be seen in figure 3.14 that as the time to expiration gets closer to zero, the curve gets closer to the option shape $(s - k)^+$.

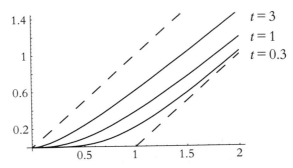

Figure 3.14 Option price against stock price for times 3, 1, and 0.3. Exercise price $k = \$1$, interest rate $r = 0$, volatility $\sigma = 1$.

Volatility dependence

All else being equal, the option is worth more the more volatile the stock is. At one extreme, if σ is very small, the option resembles a riskless bond and is just worth $(s - ke^{-rT})^+$, which is the value of the corresponding forward if the option will be in the money and is zero otherwise. At the other extreme, if σ is very large, the option is worth s.

American options

Sometimes an option has more optionality about it than just choosing between two alternatives at the maturity date. American options are the most well-known examples of such derivatives, giving the right to, say, purchase a unit of stock for a strike price k at any time up to and including the expiration date T, rather than only at that date. The buyer of the option then has to make decisions from moment to moment to decide when and if to call the option.

The buyer of an American call has the choice when to stop, and that choice can only use price information up to the present moment. Such a (random) time is called a *stopping time*. Following a strategy which will result in exercising the option at the stopping time τ, the corresponding payoff is

$$(S_\tau - k)^+ \qquad \text{at time } \tau.$$

If the option issuer knew in advance which stopping time the investor will use, the cost at time zero of hedging that payoff is

$$\mathbb{E}_\mathbb{Q}\left(e^{-r\tau}(S_\tau - k)^+\right).$$

As we do not know which τ will be used, we have to prepare for the worst possible case, and charge the maximum value (maximised over *all* possible stopping strategies),

$$V_0 = \sup_\tau \mathbb{E}_\mathbb{Q}\left(e^{-r\tau}(S_\tau - k)^+\right).$$

> **Pricing derivatives with optionality**
> In general, if the option purchaser has a set of options A, and receives a payoff X_a at time T, after choosing a in A, then the option issuer should charge
> $$V_0 = \sup_{a \in A} \mathbb{E}_\mathbb{Q}\left(e^{-rT}X_a\right)$$
> for it. If the purchaser does not exercise the option optimally, then the issuer's hedge will produce a surplus by date T.

That hedge in full

Returning to the original European option, one thing that would be useful to know would be the actual replicating strategy required, that is, to actually find out how much stock would be required at each point of time to artificially construct the derivative.

The amount of stock, ϕ_t, comes from the martingale representation theorem, but unfortunately, the theorem merely states that ϕ_t exists. However the martingale representation theorem, at heart, tells us that the reason that the discounted claim can be built from the discounted stock is that, being martingales under the same measure, one is locally just a scaled version of the other. The process ϕ_t is merely the ratio of volatilities. Thus, intuitively, if we looked at the ratio of the change in the value of the option caused by a move in the stock price and the change in the stock price used, this ought to be something like ϕ_t. And if we have a restricted enough claim where the only input required from the filtration for pricing the claim is the stock price at the current moment, and moreover that the functional relation implied by this between the value of the claim and the current stock price is smooth, then we could guess that the partial derivative of the option value with respect to the stock price is the ϕ_t we want.

And so it is. For the often-encountered case where the claim depends only on the terminal value, the option value *is* a well-behaved function of the current stock price. Suppose the derivative X is a function of the terminal value of the stock price, so that $X = f(S_T)$ for some function $f(s)$. Then the following is true.

Terminal value pricing

If the derivative X equals $f(S_T)$, for some f, then in the value of the derivative at time t is equal to $V_t = V(S_t, t)$, where $V(s, t)$ is given by the formula

$$V(s,t) = \exp\big(-r(T-t)\big)\mathbb{E}_{\mathbb{Q}}\big(f(S_T) \mid S_t = s\big)$$

And then the trading strategy is given by $\phi_t = \frac{\partial V}{\partial s}(S_t, t)$.

Why? Consider dV_t, the infinitesimal change in the value of the option. Remembering that $dS_t = \sigma S_t \, d\tilde{W}_t + rS_t \, dt$, then Itô gives us

$$dV_t = d(V(S_t, t)) = \left(\sigma S_t \frac{\partial V}{\partial s}\right) d\tilde{W}_t + \left(rS_t \frac{\partial V}{\partial s} + \tfrac{1}{2}\sigma^2 S_t^2 \frac{\partial^2 V}{\partial s^2} + \frac{\partial V}{\partial t}\right) dt.$$

But we also know that $dV_t = \phi_t \, dS_t + \psi_t \, dB_t$, from the self-financing condition. And since $dB_t = rB_t \, dt$ we have

$$dV_t = (\sigma S_t \phi_t) \, d\tilde{W}_t + (rS_t \phi_t + r\psi_t B_t) \, dt.$$

But SDE representations are unique – so the volatility terms must match, giving $\phi_t = \frac{\partial V}{\partial s}$. The amount of stock in the replicating portfolio at any stage is the derivative of the option price with respect to the stock price.

Using this substitution for ϕ_t and the fact that $V_t = S_t\phi_t + \psi_t B_t$, we can also match the drift terms of the two SDEs to get a partial differential equation for V as

$$\tfrac{1}{2}\sigma^2 s^2 \frac{\partial^2 V}{\partial s^2} + rs\frac{\partial V}{\partial s} - rV + \frac{\partial V}{\partial t} = 0.$$

Notoriously, this PDE, coupled with the boundary condition that $V(s, T)$ must equal $f(s)$, gives another way of solving the pricing equation.

Explicit Black–Scholes hedge

The call option is a terminal value claim, as described earlier, and so we can find an expression for the hedge itself. The amount of stock held is the derivative of the value function with respect to stock price. In symbols

$$\phi_t = \frac{\partial V}{\partial s}(S_t, T - t) = \Phi\left(\frac{\log \frac{S_t}{k} + (r + \frac{1}{2}\sigma^2)(T - t)}{\sigma\sqrt{T - t}}\right).$$

Because ϕ is always between zero and one, we need only ever have a bounded long position in the stock. Also the value of the bond holding at any time is

$$B_t\psi_t = -ke^{-r(T-t)}\Phi\left(\frac{\log \frac{S_t}{k} + (r - \frac{1}{2}\sigma^2)(T - t)}{\sigma\sqrt{T - t}}\right),$$

which, although always a borrowing, is bounded by the exercise price k.

There are two possibilities as the time approaches maturity. If the option is out of the money, that is the stock price is less than the exercise price, then both the bond and the stock holding go to zero, reflecting the increasing worthlessness of the option. Alternatively, if the price stays above the exercise value, then the stock holding grows to one unit and the value of the bond to $-k$. This combination exactly balances the now certain demand for a unit of stock in return for cash amount k.

Example – hedging in continuous time

This can be seen operating in practice. Below are two possible realisations of a stock price which starts at \$10. Both are exponential Brownian motions with volatility 20% and growth drift of 15%.

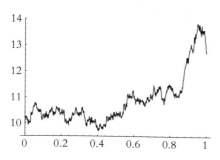

Figure 3.15a Stock price (A)

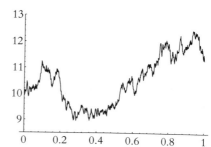

Figure 3.15b Stock price (B)

Let us price an option on this stock, to buy it at time $T = 1$ for the strike price of $k = \$12$, assuming interest rates are 5%. We can calculate both the evolving worth of the option V_t and the amount of stock to be held, ϕ_t, to hedge the contract.

In the case (A), these processes are shown in figure 3.16.

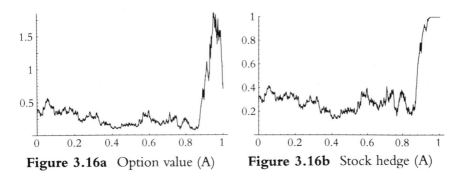

Figure 3.16a Option value (A)   **Figure 3.16b** Stock hedge (A)

As time progresses, the option becomes in the money and the option value moves like the stock price. Also the hedge gets closer and closer to one, signifying that the option will be exercised.

In the case (B), these processes are shown in Figure 3.17.

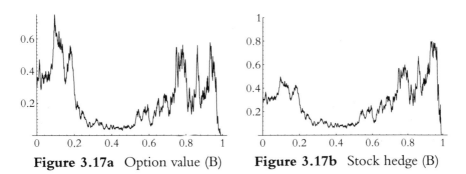

Figure 3.17a Option value (B)   **Figure 3.17b** Stock hedge (B)

This time the option is not exercised and both the value of it and the hedge go to zero over time.

 Exercise 3.16 A stock has current price $10 and moves as an exponential Brownian motion with upward drift of 15% a year (continuously compounded) and volatility of 20% a year. Current interest rates are constant at 5%. What is the value of an option on the stock for $12 in a year's time?

Exercise 3.17 For the same stock, what is the value of a derivative which pays off $1 if the stock price is more than $10 in a year's time?

Conclusions

Even with a respectable stochastic model for the stock, we can replicate any claim. Not something we had any right to expect. The replicating portfolio has a value given by the expected discounted claim, with respect to a measure which makes the discounted stock a martingale. Moreover, changing to the martingale measure has a remarkably simple effect on the process S_t – only the drift changes, to another constant value. The stock remains an exponential Brownian motion; even the volatility σ stays the same.

These three surprises conspire to make the result look easier to get at than perhaps it really is. Something subtle and beautiful really is going on under all the formalism and the result only serves to obscure it. Before we push on, stop and admire the view.

Chapter 4
Pricing market securities

The Black–Scholes model we have seen so far has a simple mathematical side but it has an even simpler financial side. The asset we considered was a stock which could be held without additional cost or benefit and was freely tradable at the price quoted. Even leaving aside the issues of transaction costs and illiquidity, not much of the financial market is like that. Even vanilla products – foreign exchange, equities and bonds – don't actually fit the simple asset class we devised. Foreign exchange involves two assets which pay interest, equities pay dividends, and bonds pay coupons.

Just retreading the same mathematics for each of these will be enough to keep us busy. The sophistication we have to peddle now is financial.

4.1 Foreign exchange

In the foreign exchange market, like the stock market, holding the basic asset, currency, is a risky business. The dollar value of, say, one pound sterling varies from moment to moment just as a US stock does. And with this risk comes demand for derivatives: claims based on the future value of one unit of currency in terms of another.

Forwards

Consider, though, a forward transaction: a dollar investor wanting to agree the cost in dollars of one pound at some future date T. As with stocks, the replicating strategy to guarantee the forward claim is static. We buy pounds now and sell dollars against them. But cash in both currencies attracts interest. And just as in the simple Black–Scholes model, our cash holding wasn't cash but a cash bond, so our cash holdings here will be cash bonds as well.

Let's make things concrete. Suppose the constant dollar interest rate is r, the sterling interest rate is u, and C_0 dollars buy a pound now. Consider the following static replicating strategy. At time t we

- own e^{-uT} units of sterling cash bonds, and

- go short $C_0 e^{-uT}$ units of dollar cash bonds.

At time zero the portfolio has nil value, and at time T the sterling holding will be one pound as required and the dollar short holding will be $C_0 e^{(r-u)T}$ – the forward price we require.

Contrast this with the stock forward price $S_0 e^{rT}$. We must be careful in extending our simple model to foreign exchange – *both* instruments now make payments. And that makes a difference.

Black–Scholes currency model

There are three instruments and processes to model – two local currency cash bonds and the exchange rate itself. Following the mathematical simplicity of Black–Scholes, our market will be:

Black–Scholes currency model
We let B_t be the dollar cash bond, D_t its sterling counterpart, and C_t be the dollar worth of one pound. Then our model is

$$\begin{aligned}
\text{Dollar bond} \quad & B_t = e^{rt}, \\
\text{Sterling bond} \quad & D_t = e^{ut}, \\
\text{Exchange rate} \quad & C_t = C_0 \exp(\sigma W_t + \mu t),
\end{aligned}$$

for some W_t a $\mathbb{P}$-Brownian motion and constants r, u, σ and μ.

The dollar investor

The underlying finance dictates that there are two tradables available to the dollar investor. One is uncomplicated – the dollar bond is straightforwardly a dollar tradable much as the cash bond was in the basic account of Black–Scholes. But the other is not.

We would like to think of the stochastic process C_t, the exchange rate, as a tradable but it isn't. The process C_t represents the dollar value of one pound sterling, but sterling cash isn't a tradable instrument in our market. To hold cash naked would be to set up an arbitrage against the cash bond – to put it another way, the existence of the sterling cash bond D_t sets an interest rate for sterling cash by arbitrage, and that rate is u not zero.

On the other hand, D_t by itself isn't a dollar tradable either – it is the price of a tradable instrument, but it's a *sterling* price.

Fortunately, the product of the two $S_t = C_t D_t$ *is a dollar tradable*. The dollar investor can hold sterling cash bonds, and the dollar value of the holding will be given by the translation of the sterling price D_t into dollars, that is by multiplication by C_t.

Translation, then, yields two processes, B_t and S_t, which mirror the basic Black–Scholes set up.

Three steps to replication (foreign exchange)

(1) Find a measure $\mathbb{Q}$ under which the sterling bond discounted by the dollar bond $Z_t = B_t^{-1} S_t = B_t^{-1} C_t D_t$ is a martingale.

(2) Form the process $E_t = \mathbb{E}_\mathbb{Q}\left(B_T^{-1} X \mid \mathcal{F}_t\right)$.

(3) Find a previsible process ϕ_t, such that $dE_t = \phi_t \, dZ_t$.

Step one

The dollar discounted worth of the sterling bond is

$$Z_t = C_0 \exp\left(\sigma W_t + (\mu + u - r)t\right).$$

Can we make this into a martingale under some new measure $\mathbb{Q}$? Only if $\tilde{W}_t = W_t + \sigma^{-1}(\mu + u - r + \tfrac{1}{2}\sigma^2)t$ is a $\mathbb{Q}$-Brownian motion, which is made possible as before by the Cameron–Martin–Girsanov theorem. Then, under $\mathbb{Q}$

$$Z_t = C_0 \exp\left(\sigma \tilde{W}_t - \tfrac{1}{2}\sigma^2 t\right),$$

and thus $\qquad C_t = C_0 \exp\left(\sigma \tilde{W}_t + (r - u - \tfrac{1}{2}\sigma^2)t\right).$

Step two

Given this $\mathbb{Q}$, define the process E_t to be the conditional expectation process $\mathbb{E}_\mathbb{Q}(B_T^{-1} X | \mathcal{F}_t)$, which as noted before is a $\mathbb{Q}$-martingale.

Step three

The martingale representation theorem produces an $\mathcal{F}$-previsible process ϕ_t linking E_t with Z_t, such that

$$E_t = E_0 + \int_0^t \phi_s \, dZ_s.$$

Now where? We need a replicating strategy (ϕ_t, ψ_t) detailing holdings of our two dollar tradables S_t and B_t, so we try

- holding ϕ_t units of sterling cash bond, and

- holding $\psi_t = E_t - \phi_t Z_t$ units of dollar cash bond.

The dollar value of the replicating portfolio at time t is $V_t = \phi_t S_t + \psi_t B_t = B_t E_t$. This portfolio is only self-financing if changes in its value are only due to changes in the assets' prices, that is $dV_t = \phi_t \, dS_t + \psi_t \, dB_t$, or as was shown to be equivalent in section 3.7, if $dE_t = \phi_t \, dZ_t$ – which is precisely what the martingale representation theorem guarantees.

Since $V_T = B_T E_T$, and E_T is the discounted claim $B_T^{-1} X$, we have a self-financing strategy (ϕ_t, ψ_t) which replicates our arbitrary claim X.

> **Option price formula (foreign exchange)**
> All claims have arbitrage prices and those prices are given by the portfolio value
> $$V_t = B_t \, \mathbb{E}_\mathbb{Q}(B_T^{-1} X \mid \mathcal{F}_t).$$
> where $\mathbb{Q}$ is the measure under which the discounted asset Z_t is a martingale.

Example – forward contract

A sterling forward contract. At what price should we agree to trade sterling at a future date T? If we agree to buy a unit of sterling for an amount k of dollars, our payoff at time T is

$$X = C_T - k.$$

Its worth at time t is $V_t = B_t \mathbb{E}_{\mathbb{Q}}(B_T^{-1} X | \mathcal{F}_t)$ which is $e^{-r(T-t)} \mathbb{E}_{\mathbb{Q}}(C_T - k | \mathcal{F}_t)$. So the forward price at time zero for purchasing sterling at time T is $k = \mathbb{E}_{\mathbb{Q}}(C_T)$ or

$$F = \mathbb{E}_{\mathbb{Q}}\left(C_0 \exp\left(\sigma \tilde{W}_T + (r - u - \tfrac{1}{2}\sigma^2)T\right)\right) = e^{(r-u)T} C_0.$$

That is, the current price for sterling discounted by a factor depending on the difference between the interest rates of the two currencies. With this strike, the contract's value at time t is

$$V_t = e^{-uT}\left(e^{ut}C_t - e^{rt}C_0\right).$$

The discounted portfolio value is $E_t = B_t^{-1}V_t = e^{-uT}Z_t - e^{-uT}C_0$, thus $dE_t = e^{-uT}\, dZ_t$, and so the required hedge ϕ_t is the constant e^{-uT}, and ψ_t is the constant $-e^{-uT}C_0$.

This confirms our earlier intuition.

Example – call option

A sterling call. Suppose we have a contract which allows us the option of buying a pound at time T in the future for the price of k dollars. The dollar payoff at time T is

$$X = (C_T - k)^+.$$

The value of the payoff at time t is $V_t = B_t \mathbb{E}_{\mathbb{Q}}(B_T^{-1} X | \mathcal{F}_t)$. Because C_T is log-normally distributed we can evaluate this easily using a probabilistic result:

> **Log-normal call formula**
> If Z is a normal $N(0,1)$ random variable, and F, $\bar{\sigma}$ and k are constants, then
>
> $$\mathbb{E}\left((F\exp(\bar{\sigma}Z - \tfrac{1}{2}\bar{\sigma}^2) - k)^+\right) = F\Phi\left(\frac{\log\frac{F}{k} + \tfrac{1}{2}\bar{\sigma}^2}{\bar{\sigma}}\right) - k\Phi\left(\frac{\log\frac{F}{k} - \tfrac{1}{2}\bar{\sigma}^2}{\bar{\sigma}}\right).$$

As the forward price F is $\mathbb{E}_\mathbb{Q}(C_T)$, the value of C_T can be written in the form $F\exp(\bar\sigma Z - \frac{1}{2}\bar\sigma^2)$, where $\bar\sigma^2$ is the variance of $\log C_T$, namely $\sigma^2 T$, and Z is a normal $N(0,1)$ under $\mathbb{Q}$.

The option price at time zero is then $\mathbb{E}\big((F\exp(\bar\sigma Z - \frac{1}{2}\bar\sigma^2) - k)^+\big)$, which the theorem tells us is

$$V_0 = e^{-rT}\left\{ F\Phi\left(\frac{\log\frac{F}{k} + \frac{1}{2}\sigma^2 T}{\sigma\sqrt{T}}\right) - k\Phi\left(\frac{\log\frac{F}{k} - \frac{1}{2}\sigma^2 T}{\sigma\sqrt{T}}\right)\right\}.$$

The hedge is

$$\phi_t = e^{-uT}\Phi\left(\frac{\log\frac{F_t}{k} + \frac{1}{2}\sigma^2(T-t)}{\sigma\sqrt{T-t}}\right),$$

$$\psi_t = -ke^{-rT}\Phi\left(\frac{\log\frac{F_t}{k} - \frac{1}{2}\sigma^2(T-t)}{\sigma\sqrt{T-t}}\right),$$

where F_t is the forward sterling price at time t, $F_t = e^{(r-u)(T-t)}C_t$.

The sterling investor

A sterling investor sees things differently. Were we operating in pounds we would not be wanting dollar price processes of tradable instruments but sterling ones. The first of these is simply the sterling bond $D_t = e^{ut}$, which will be our basic unit of account. There is also the inverse exchange rate process C_t^{-1} – the worth in pounds of one dollar. This has the value

$$C_t^{-1} = C_0^{-1}\exp(-\sigma W_t - \mu t),$$

but it is not the sterling price of a tradable instrument, any more than C_t was for the dollar investor. Our other actual sterling tradable price process is the sterling value of the *dollar bond* $C_t^{-1}B_t$.

With our two sterling tradable prices, D_t and $C_t^{-1}B_t$, we can follow again our three-step replication programme. The sterling discounted value of the dollar bond is

$$Y_t = D_t^{-1}C_t^{-1}B_t = C_0^{-1}\exp\big(-\sigma W_t - (\mu + u - r)t\big).$$

This discounted price process Y_t will be a martingale under the new measure $\mathbb{Q}^\mathcal{L}$, if

$$\tilde W_t^\mathcal{L} = W_t + \sigma^{-1}(\mu + u - r - \tfrac{1}{2}\sigma^2)t$$

is $\mathbb{Q}^{\mathcal{L}}$-Brownian motion. Then hedging will be possible as before.

> **Option price formula (sterling investor)**
> The value to the sterling investor of a sterling payoff X at time T is
>
> $$U_t = D_t \, \mathbb{E}_{\mathbb{Q}^{\mathcal{L}}} (D_T^{-1} X \mid \mathcal{F}_t).$$
>
> where $\mathbb{Q}^{\mathcal{L}}$ is the measure under which the sterling discounted asset Y_t is a martingale.

Change of numeraire

A worrying possibility now surfaces – the measures $\mathbb{Q}$ and $\mathbb{Q}^{\mathcal{L}}$ are different. Will the dollar and sterling investors disagree about the price of the same security?

Suppose X is a dollar claim which pays off at time T. To the dollar investor, the claim is worth at time t

$$V_t = B_t \, \mathbb{E}_{\mathbb{Q}} (B_T^{-1} X \mid \mathcal{F}_t) \quad \text{dollars.}$$

To the sterling investor, the claim pays off $C_T^{-1} X$ pounds, rather than X dollars, at time T, and its sterling worth at time t is

$$U_t = D_t \, \mathbb{E}_{\mathbb{Q}^{\mathcal{L}}} (D_T^{-1}(C_T^{-1} X) \mid \mathcal{F}_t) \quad \text{pounds.}$$

Do these two prices agree? That is, is the dollar worth of the sterling valuation, $C_t U_t$, the same as the original dollar valuation V_t?

The $\mathbb{Q}^{\mathcal{L}}$-Brownian motion $\tilde{W}_t^{\mathcal{L}}$ is equal to $\tilde{W}_t - \sigma t$, so that by the converse of the Cameron–Martin–Girsanov theorem the Radon–Nikodym derivative of $\mathbb{Q}^{\mathcal{L}}$ with respect to $\mathbb{Q}$ (up to time T) must be

$$\frac{d\mathbb{Q}^{\mathcal{L}}}{d\mathbb{Q}} = \exp\left(\sigma \tilde{W}_T - \tfrac{1}{2}\sigma^2 T\right).$$

The $\mathbb{Q}$-martingale associated with the Radon–Nikodym derivative, formed by conditional expectation is

$$\zeta_t = \mathbb{E}_{\mathbb{Q}}\left(\frac{d\mathbb{Q}^{\mathcal{L}}}{d\mathbb{Q}} \,\Big|\, \mathcal{F}_t\right) = \exp(\sigma \tilde{W}_t - \tfrac{1}{2}\sigma^2 t).$$

Note that ζ_t is (up to a constant) the dollar discounted worth of the sterling bond. Concretely, $C_0\zeta_t = Z_t = B_t^{-1}C_tD_t$. Recall also (Radon–Nikodym fact (ii) of section 3.4) that for any random variable X which is known by time T,

$$\mathbb{E}_{\mathbb{Q}\mathcal{L}}(X|\mathcal{F}_t) = \zeta_t^{-1}\mathbb{E}_{\mathbb{Q}}(\zeta_T X|\mathcal{F}_t).$$

So the dollar worth of the sterling investor's valuation is

$$C_tU_t = C_tD_t\mathbb{E}_{\mathbb{Q}\mathcal{L}}\left(D_T^{-1}C_T^{-1}X \mid \mathcal{F}_t\right) = C_tD_t\zeta_t^{-1}\mathbb{E}_{\mathbb{Q}}\left(\zeta_T D_T^{-1}C_T^{-1}X \mid \mathcal{F}_t\right),$$

which is (substituting in the ζ_t expression) equal to

$$C_tU_t = B_t\,\mathbb{E}_{\mathbb{Q}}\left(B_T^{-1}X \mid \mathcal{F}_t\right) = V_t.$$

Thus the payoff of X dollars at time T is worth the same to either investor at any time beforehand. Similar calculations show that the dollar and sterling investors' replicating strategies for X are identical. So they agree not only on the prices but also on the hedging strategy.

The difference of martingale measures only reflected the different numeraires of the two investors rather than any fundamental disagreement over prices. Further details on the effect, or lack of it, of changing numeraires are in section 6.4.

All investors, whatever their currency of account, will agree on the current value of a derivative or other security.

4.2 Equities and dividends

An equity is a stock which makes periodic cash payments to the current holder. Our previous models treated a stock as a pure asset, but they can be modified to handle dividend payments.

It is simplest to begin with a dividend which is paid continuously.

Equity model with continuous dividends

Let the stock price S_t follow a Black–Scholes model, $S_t = S_0 \exp(\sigma W_t + \mu t)$ and B_t be a constant-rate cash bond $B_t = \exp(rt)$. The dividend payment made in the time interval of length dt starting at time t is

$$\delta S_t \, dt,$$

where δ is a constant of proportionality.

Just as with foreign exchange, our problem is that the process S_t is not a tradable asset. If we buy the stock for S_0, by the time we come to sell it at time t, what we bought is worth not just the price of the stock itself, namely S_t, but also the total accumulated dividends, which under the model will depend on all the different values that the stock has taken up until time t. The process S_t is no longer the value of the asset as a whole, because it is not enough.

We need to translate S_t somehow, and to find a new process as we did in foreign exchange, which involves S_t but *is* a tradable. Consider the following simple portfolio strategy. The portfolio starts with one unit of stock, costing S_0, and at every instant when the cash dividend is paid out, that cash is immediately used to buy a little more stock. That is, we are continuously reinvesting the dividends in the stock. The infinitesimal payout is $\delta S_t \, dt$ per unit of stock, which will purchase $\delta \, dt$ more units of stock. At time t, the number of stock units held by the portfolio will be $\exp(\delta t)$, and the worth of the portfolio is

$$\tilde{S}_t = S_0 \exp\big(\sigma W_t + (\mu + \delta)t\big).$$

Note how the structure of the model's assumptions made the translation straightforward. We assumed that the dividend payments were a constant proportion of the stock price. As a consequence it made it natural to construct the tradable by reinvesting in the stock. If we had assumed that the dividend stream was known in advance, independent of the stock price, then we would have reinvested in the cash bond (for an example of this see section 4.3 on bonds). Assumptions are all.

107

Replicating strategies – equities

Our definition of a portfolio of stock and bond (ϕ_t, ψ_t) can be rewritten as a portfolio of the reinvested stock and bond $(\tilde{\phi}_t, \psi_t)$, where $\tilde{\phi}_t = e^{-\delta t}\phi_t$, with value $V_t = \phi_t S_t + \psi_t B_t = \tilde{\phi}_t \tilde{S}_t + \psi_t B_t$. The advantage of the new framework is that the self-financing equation retains the familiar form

$$dV_t = \tilde{\phi}_t \, d\tilde{S}_t + \psi_t \, dB_t,$$

whereas in the plain stock/bond notation, this equation would need to be modified by the dividend cash stream, becoming $dV_t = \phi_t \, dS_t + \psi_t \, dB_t + \phi_t \delta S_t \, dt$. That is, changes in the portfolio value are due both to trading profits and losses (the dS_t and dB_t terms) and also to dividend payments.

Working now with our reinvested stock, as usual we want to make the discounted asset $\tilde{Z}_t = B_t^{-1}\tilde{S}_t$ into a martingale. Now $\tilde{Z}_t$ has SDE

$$d\tilde{Z}_t = \tilde{Z}_t\big(\sigma \, dW_t + (\mu + \delta + \tfrac{1}{2}\sigma^2 - r) \, dt\big),$$

so that we want a measure $\mathbb{Q}$ under which $\tilde{W}_t = W_t + \sigma^{-1}(\mu + \delta + \tfrac{1}{2}\sigma^2 - r)t$ is Brownian motion. So under this martingale measure $\mathbb{Q}$, $d\tilde{Z}_t = \sigma\tilde{Z}_t \, d\tilde{W}_t$. To construct a strategy to hedge a claim X maturing at date T, again we follow the simple Black–Scholes model, and use the martingale representation theorem. That is, there exists a previsible process $\tilde{\phi}_t$ such that

$$E_t = \mathbb{E}_{\mathbb{Q}}\big(B_T^{-1}X \mid \mathcal{F}_t\big) = \mathbb{E}_{\mathbb{Q}}\big(B_T^{-1}X\big) + \int_0^t \tilde{\phi}_s \, d\tilde{Z}_s.$$

The trading strategy is to hold $\tilde{\phi}_t$ units of the translated asset $\tilde{S}_t$ and $\psi_t = E_t - \tilde{\phi}_t\tilde{Z}_t$ units of the cash bond. In terms of our original securities, this amounts to holding $\phi_t = e^{\delta t}\tilde{\phi}_t$ units of the stock S_t and the same ψ_t units of the bond B_t.

Thus, under the martingale measure

$$S_t = S_0 \exp\big(\sigma\tilde{W}_t + (r - \delta - \tfrac{1}{2}\sigma^2)t\big),$$

which is log-normally distributed.

Example – forward

An agreement to buy a unit of stock at time T for amount k has payoff

$$X = S_T - k.$$

Its worth at time t is

$$V_t = \mathbb{E}_{\mathbb{Q}}\big(e^{-r(T-t)}(S_T - k) \mid \mathcal{F}_t\big) = e^{-\delta(T-t)}S_t - e^{-r(T-t)}k.$$

The value of k which gives the contract initial nil value is the forward price of S_T,

$$F = e^{(r-\delta)T}S_0.$$

The hedge is then to hold $\phi_t = e^{-\delta(T-t)}$ units of the stock and $\psi_t = -ke^{-rT}$ units of the bond at time t. Note the slightly surprising dynamic strategy for the forward. Instead of simply holding a certain amount of stock until T, we are continually buying more with the dividend income. Why? Again because of our assumption – if the dividend payments are a known proportion of the stochastic S_t, we have no choice but to hide them in the stock itself.

Example – call option

A call struck at k, exercised at time T has payoff $X = (S_T - k)^+$, and value at time zero of $V_0 = \mathbb{E}_{\mathbb{Q}}\big(e^{-rT}(S_T - k)^+\big)$, which equals

$$V_0 = e^{-rT}\left\{ F\Phi\left(\frac{\log\frac{F}{k} + \frac{1}{2}\sigma^2 T}{\sigma\sqrt{T}}\right) - k\Phi\left(\frac{\log\frac{F}{k} - \frac{1}{2}\sigma^2 T}{\sigma\sqrt{T}}\right)\right\},$$

where F is the forward price above $e^{(r-\delta)T}S_0$. The hedge will be to hold $e^{-\delta(T-t)}\Phi(+)$ units of the stock and have a negative holding of $ke^{-rT}\Phi(-)$ units of the bond. (Here $\Phi(+)$ and $\Phi(-)$ refer respectively to the two Φ terms in the above equation.)

Again the Black–Scholes call option formula re-emerges – if the martingale measure $\mathbb{Q}$ makes the process under study, S_t, have a log-normal distribution, then the theorem in section 4.1 comes into play. Knowing the forward F and the term volatility σ is enough to specify the price.

Example – guaranteed equity profits

A contract pays off according to gains of the UK FTSE stock index S_t, with a guaranteed minimum payout and a maximum payout. More precisely, it is a five-year contract which pays out 90% times the ratio of the terminal and initial values of FTSE. Or it pays out 130% if otherwise it would be less, or 180% if otherwise it would be more. How much is this payout worth?

Our data are

$$\text{FTSE drift} \quad \mu = 7\%$$
$$\text{FTSE volatility} \quad \sigma = 15\%$$
$$\text{FTSE dividend yield} \quad \delta = 4\%$$
$$\text{UK interest rate} \quad r = 6.5\%$$

As FTSE is composed of 100 different stocks, their separate dividend payments will approximate a continuously paying stream. The claim X is

$$X = \min\{\max\{1.3, 0.9S_T\}, 1.8\},$$

where T is 5 years and the initial FTSE value S_0 is 1. This claim can be rewritten as

$$X = 1.3 + 0.9\{(S_T - 1.444)^+ - (S_T - 2)^+\}.$$

That is, X is actually the difference of two FTSE calls (plus some cash). The forward price for S_T is

$$F = e^{(r-\delta)T}S_0 = 1.133.$$

Using the above call price formula for dividend-paying stocks, we can value these calls (per unit) at 0.0422 and 0.0067 respectively. The worth of X at time zero is then

$$V_0 = 1.3e^{-rT} + 0.9(0.0422 - 0.0067) = 0.9712.$$

Were we to have forgotten that the constituent stocks of FTSE pay dividends, but the dividends are not reflected in the index, we would incorrectly have valued the contract at 1.0183 – about 5% too high.

Periodic dividends

In practice, an individual stock pays dividends at regular intervals rather than continuously, but this presents no real problems for our basic model. Let us assume that the times of dividend payments T_1, T_2, ... are known in advance, and at each time T_i, the current holder of the equity receives a payment of a fraction δ of the current stock price. The stock price must also instantaneously decrease by the same amount – or else there would be an arbitrage opportunity. At any time $T = T_i$, then, we can assume the dividend payout exactly equals the instantaneous decrease in the stock price.

Equity model with periodic dividends

At deterministic times T_1, T_2, ..., the equity pays a dividend of a fraction δ of the stock price which was current just before the dividend is paid. The stock price process itself is modelled as

$$S_t = S_0(1 - \delta)^{n[t]} \exp(\sigma W_t + \mu t),$$

where $n[t] = \max\{i : T_i \leqslant t\}$ is the number of dividend payments made by time t. There is also a cash bond $B_t = \exp(rt)$.

We face two problems. The first is the familiar one that S_t is not by itself the price of tradable asset. Translation, however, should provide a cure. The second is more serious. Away from the times T_i, S_t has the usual SDE of $dS_t = S_t(\sigma \, dW_t + (\mu + \tfrac{1}{2}\sigma^2) \, dt)$, but at those times it has discontinuous jumps. Thus S_t is discontinuous – it doesn't fit our definition of a stochastic process. Fortunately, translation cures this as well.

Consider the following trading strategy. Starting with one unit of stock, every time the stock pays a dividend we reinvest the dividend by buying more stock. At time t, we will have $(1 - \delta)^{-n[t]}$ units of the stock, and the value of our portfolio will be $\tilde{S}_t$, where

$$\tilde{S}_t = (1 - \delta)^{-n[t]} S_t = S_0 \exp(\sigma W_t + \mu t).$$

As before, $\tilde{S}_t$ is tradable but our arbitrage justified assumption that the dividend payments match the stock price jumps feeds through into making $\tilde{S}_t$ continuous as well. We are back in familiar territory.

Replicating strategy

Our trading strategy will then be $(\tilde{\phi}_t, \psi_t)$, where $\tilde{\phi}_t$ is the number of units of $\tilde{S}_t$ we hold at time t, and ψ_t is the amount of the cash bond B_t. Such a strategy is equivalent to holding $\phi_t = (1 - \delta)^{-n[t]}\tilde{\phi}_t$ units of the actual stock S_t.

The discounted value of the $(\tilde{\phi}_t, \psi_t)$ portfolio is $E_t = \tilde{\phi}_t\tilde{Z}_t + \psi_t$, where $\tilde{Z}_t$ is the discounted value of the reinvested stock price $\tilde{Z}_t = B_t^{-1}\tilde{S}_t$. The portfolio will be self-financing if $dE_t = \tilde{\phi}_t\, d\tilde{Z}_t$.

As before, we want to find a $\mathbb{Q}$ which makes $\tilde{Z}_t$ into a martingale. As $d\tilde{Z}_t = \tilde{Z}_t(\sigma\, dW_t + (\mu - r)\, dt)$, this will have no drift if $\tilde{W}_t = W_t + \sigma^{-1}(\mu + \tfrac{1}{2}\sigma^2 - r)t$ is $\mathbb{Q}$-Brownian motion. Then $\tilde{Z}_t$ is also a $\mathbb{Q}$-martingale.

We can form the process $E_t = \mathbb{E}_{\mathbb{Q}}(B_T^{-1}X|\mathcal{F}_t)$, where X is the option on the stock which we wish to hedge.

Finally, the martingale representation theorem produces a hedging process $\tilde{\phi}_t$ and the corresponding ψ_t can be set to be $E_t - \tilde{\phi}_t\tilde{Z}_t$. So hedging is still possible in this case, and the value at time zero of the claim X is $\mathbb{E}_{\mathbb{Q}}(B_T^{-1}X)$.

The stock price, under $\mathbb{Q}$, is

$$S_t = S_0(1-\delta)^{n[t]}e^{\sigma\tilde{W}_t + (r - \frac{1}{2}\sigma^2)t}.$$

Since this is log-normal, with the forward price for S_T equal to $F = S_0(1 - \delta)^{n[T]}e^{rT}$, the Black–Scholes price for a call option struck at k is equal to

$$V_0 = e^{-rT}\left\{F\Phi\left(\frac{\log\frac{F}{k} + \frac{1}{2}\sigma^2 T}{\sigma\sqrt{T}}\right) - k\Phi\left(\frac{\log\frac{F}{k} - \frac{1}{2}\sigma^2 T}{\sigma\sqrt{T}}\right)\right\}.$$

4.3 Bonds

A pure discount bond is a security which pays off one unit at some future maturity time T. Were interest rates completely constant at rate r it would have present value at time t of $e^{-r(T-t)}$. We might, however, want to consider the effect of interest rates being stochastic – much as they are in real markets. And with varying interest rates, uncertainty about their future values would cause a discount bond price to move randomly as well.

A full model of discount bonds, or for that matter coupon bonds, will have to wait for chapter five and term structure models. The interplay of interest rates of different maturities and the arbitrage minefield that models have to tiptoe through is not something we want to worry about in a simple Black–Scholes account. As a consequence we will try to take a schizophrenic attitude to interest rates. Bond prices will vary stochastically, but the short-term interest rate will be deterministic. In the real markets there is clearly a link, but then it can be argued that there are links between stock or foreign exchange prices and the cash bonds as well. Over short time horizons most practitioners ignore these links in all three markets.

Discount bonds

The Black–Scholes model for discount bonds is:

Discount bond model

We assume a cash bond $B_t = \exp(rt)$ for some positive constant r, and a discount bond S_t whose price follows

$$S_t = S_0 \exp(\sigma W_t + \mu t),$$

for all times t less than T, some time horizon T long before the maturity time τ of the bond.

In formulation, this model is indistinguishable from the simple Black–Scholes model for stocks. Thus the forward price for purchasing the bond at time $T < \tau$ is

$$F = \mathbb{E}_\mathbb{Q}(S_T),$$

where $\mathbb{Q}$ is the measure under which $e^{-rt}S_t$ is a martingale. Since $\sigma^2 T$ is the variance, under $\mathbb{Q}$, of $\log S_T$ (σ is the term volatility), then the price of a call on S_T struck at k is

$$e^{-rT}\left\{ F\Phi\left(\frac{\log \frac{F}{k} + \frac{1}{2}\sigma^2 T}{\sigma\sqrt{T}} \right) - k\Phi\left(\frac{\log \frac{F}{k} - \frac{1}{2}\sigma^2 T}{\sigma\sqrt{T}} \right) \right\}.$$

We have to be careful, though, with our assumption that T is much before the maturity τ. Not only does the distinction between the deterministic cash bond and the stochastic discount bond get harder to maintain as T approaches τ, but for similar reasons it gets harder to justify a simple drift μ and a constant positive σ. The bond promises one unit at time τ, thus its price at time τ must be $S_\tau = 1$. In a good model, the drift and volatility will conspire to ensure this *pull to par* – and indeed this will happen in chapter five. Here if we let $T = \tau$, we would have no such guarantee.

Bonds with coupons

Most market bonds do not just pay off one unit at maturity, but also pay off a series of smaller amounts c at various pre-determined times $T_1, T_2, \ldots, T_n$ before maturity. Such coupon payments may resemble dividend payments, but unlike the equity model, the amount of the coupon is known in advance. Here the schizophrenia extends to the treatment of coupons before and after the expiry date, T, of the option. The simplest model is to view coupons that occur before time T as coming under the regime of the deterministic cash bond, and coupons occurring after time T (including the redemption payment at maturity) as following a stochastic price process.

Coupon bond model

There is a simple cash bond $B_t = \exp(rt)$, and a coupon bond which pays off an amount c at times T_1, T_2, ..., up to a horizon τ. Denoting $I(t) = \min\{i : t < T_i\}$ to be the sequence number of the next coupon payment after time t, and j to be $I(T) - 1$, the total number of payments before time T, then the price of the bond at time t is then

$$S_t = \sum_{i=I(t)}^{j} ce^{-r(T_i - t)} + A\exp(\sigma W_t + \mu t), \qquad t < T.$$

Specifically, we model the first sort of coupon (payable at, for example, $T_i < T$) to be worth

$$ce^{-r(T_i - t)} \qquad \text{at time } t \ (t < T_i),$$

and for the sum of all the post-T payments to evolve as an exponential

Brownian motion

$$A \exp(\sigma W_t + \mu t), \qquad \text{for } t < T,$$

for constants A, σ, and μ.

Again S_t is discontinuous at the coupon payment times, and again we can use a translation rather like the one used for equity dividends (section 4.2). But because the coupon payments are known in advance, this time we manufacture a continuous tradable asset by holding one unit of the coupon bond and investing all the coupon payments, as they occur, in the *cash bond*. The value of this asset is $\tilde{S}_t$, where

$$\tilde{S}_t = \sum_{i=1}^{j} ce^{-r(T_i - t)} + A \exp(\sigma W_t + \mu t).$$

This is now a tradable asset with a continuous stochastic process.

Replicating strategy

We describe a portfolio as $(\tilde{\phi}_t, \psi_t)$, where $\tilde{\phi}_t$ is the amount of the asset $\tilde{S}_t$ held at time t, and ψ_t is the direct holding of the cash bond $B_t = e^{rt}$. We let V_t be the value of the portfolio, $V_t = \tilde{\phi}_t \tilde{S}_t + \psi_t B_t$, and E_t be its discounted value $E_t = \tilde{\phi}_t \tilde{Z}_t + \psi_t$, where $\tilde{Z}_t$ is the discounted value of the asset $\tilde{S}_t$. The portfolio is self-financing if $dE_t = \tilde{\phi}_t \, d\tilde{Z}_t$.

As usual, we want to make $\tilde{Z}_t$ into a martingale by changing measure. In fact $\tilde{Z}_t$ is just a constant cash sum of $\sum_{i=1}^{j} ce^{-rT_i}$ plus an exponential Brownian motion $A \exp(\sigma W_t + (\mu - r)t)$. This will be a $\mathbb{Q}$-martingale if $\tilde{W}_t = W_t + \sigma^{-1}(\mu + \tfrac{1}{2}\sigma^2 - r)t$ is $\mathbb{Q}$-Brownian motion.

For an option X payable at time T, the process $E_t = \mathbb{E}_{\mathbb{Q}}(B_T^{-1} X | \mathcal{F}_t)$ can be represented as $dE_t = \tilde{\phi}_t \, d\tilde{Z}_t$ for some previsible process $\tilde{\phi}_t$. We can set ψ_t to be $E_t - \tilde{\phi}_t \tilde{Z}_t$, so that $(\tilde{\phi}_t, \psi_t)$ is a hedging strategy for X. The value of X at time zero must now be $\mathbb{E}_{\mathbb{Q}}(B_T^{-1} X)$.

Under $\mathbb{Q}$, the price of the bond at time T is just

$$S_T = A \exp(\sigma \tilde{W}_T + (r - \tfrac{1}{2}\sigma^2)T).$$

This is log-normally distributed, so we can follow the call formula from section 4.1 to see that the forward price for S_T is $F = Ae^{rT}$ and the value of a call on S_T struck at k is

$$e^{-rT} \left\{ F\Phi\left(\frac{\log \frac{F}{k} + \tfrac{1}{2}\sigma^2 T}{\sigma\sqrt{T}} \right) - k\Phi\left(\frac{\log \frac{F}{k} - \tfrac{1}{2}\sigma^2 T}{\sigma\sqrt{T}} \right) \right\}.$$

4.4 Market price of risk

Now is the time to tie some loose ends together. The same pattern has been repeating through all the examples so far – the stochastic processes we have been using as models in this chapter have been tied to tradable quantities only indirectly. The foreign exchange process had to be converted from a non-tradable cash process to a tradable discount bond process. For equities, the model process had to have dividends recombined to make it tradable. And for bonds, the coupons had to be reinvested in the numeraire process. Underlying all this was a tradable/non-tradable distinction – we couldn't use the martingale representation theorem to replicate claims until we had something tradable to replicate with. But the distinction has so far been a common sense one – can we do any better?

To some extent, yes. Some of the tradable/non-tradable distinction is going to have to be founded on goodwill. After all whether something can be traded or not in a free market is not a mathematical decision. But if we decide on a particular process S_t representing something truly tradable and select an appropriate discounting process B_t, then we can explore the market they create.

Martingales are tradables

Suppose that there is some measure $\mathbb{Q}$ under which the discounted tradable, $Z_t = B_t^{-1} S_t$, is a $\mathbb{Q}$-martingale, what can we say about another process V_t adapted to the same filtration $\mathcal{F}_t$ such that $E_t = B_t^{-1} V_t$ is also a $\mathbb{Q}$-martingale?

Firstly, the martingale representation theorem gives us that, as long as Z_t has non-zero volatility, we can find an $\mathcal{F}$-previsible process ϕ_t such that

$$dE_t = \phi_t \, dZ_t.$$

Taking our cue from all the examples so far, we could create a portfolio (ϕ_t, ψ_t) where at time t we are

- long ϕ_t of the tradable S_t,

- long $\psi_t = E_t - \phi_t Z_t$ of the tradable B_t.

Then as before we can show that (ϕ_t, ψ_t) is a self-financing strategy, that is changes in the value of the (ϕ_t, ψ_t) portfolio are explainable in terms of changes in value of the tradable constituents alone. And the value of this portfolio at time t is always exactly V_t.

In other words we can make V_t out of S_t and B_t. So it seems reasonable enough to ennoble V_t with the title tradable as well. Being a $\mathbb{Q}$-martingale after discounting is enough to ensure that it can be made costlessly from tradables – so it might as well be tradable itself. Of course all the derivatives that we have been constructing out of claims have this property – $\mathbb{E}_{\mathbb{Q}}(B_T^{-1}X|\mathcal{F}_t)$ is always a $\mathbb{Q}$-martingale.

Non-martingales are non-tradables

What about the other way round? Suppose $B_t^{-1}V_t$ was not a $\mathbb{Q}$-martingale. Then from our definition of a martingale, there must be a positive probability at some times T and s that $\mathbb{E}_{\mathbb{Q}}(B_T^{-1}V_T|\mathcal{F}_s) \neq B_s^{-1}V_s$. What would happen if V_t were tradable and the market stumbled into this possible filtration?

Suppose we define another process U_t by simply setting U_t to be the cost of replicating the claim V_T, that is $U_t = B_t\,\mathbb{E}_{\mathbb{Q}}(B_T^{-1}V_T|\mathcal{F}_t)$. Then the terminal value of U_T will be equal to V_T but at time s, U_s and V_s will be (possibly) different. As $B_t^{-1}U_t$ is a $\mathbb{Q}$-martingale we can view U_t as tradable by dint of being able to construct it from S_t and B_t.

So we have two tradables, U_t and V_t, such that they are identical at time T but different at some earlier time s (with positive probability). We then have an arbitrage engine. If, say, U_s were greater than V_s, we could buy unlimited amounts of V and sell unlimited amounts of U collecting the cash up front. The $V - U$ portfolio can be sold for nothing at time T, leaving just the (invested) cash as a guaranteed profit. And if U_s were less than V_s we would run the engine in reverse.

Thus if V_t were genuinely tradable, the market formed by S_t, B_t and V_t would contain arbitrage opportunities – something we might want to dismiss by *fiat*. To avoid arbitrage engines, then, if $B_t^{-1}V_t$ were not a $\mathbb{Q}$-martingale, it had better not be tradable.

We have something akin to a definition then. Within an established (complete) market of tradable securities, there is a straightforward way of checking whether another process is a tradable security or not. It is tradable if its discounted price is a martingale under the martingale measure $\mathbb{Q}$, and is not tradable if it isn't.

Tradable securities

Given a numeraire B_t and a tradable asset S_t, a process V_t represents a tradable asset if and only if its discounted value $B_t^{-1}V_t$ is actually a $\mathbb{Q}$-martingale, where $\mathbb{Q}$ is the measure under which the discounted asset, $B_t^{-1}S_t$, is a martingale.

One way round, the process is just part of the 'linear span' of S_t and B_t; the other way round, there is only room for two 'independent' tradables in a market defined by one-dimensional Brownian motion – any more and there can be arbitrage.

 Exercise 4.1 If S_t is a tradable Black–Scholes stock price under the martingale measure $\mathbb{Q}$, $S_t = \exp\left(\sigma \tilde{W}_t + (r - \frac{1}{2}\sigma^2)t\right)$, with cash bond $B_t = \exp(rt)$, show that
(i) $X_t = S_t^2$ is non-tradable,
(ii) $X_t = S_t^{-\alpha}$, where $\alpha = 2r/\sigma^2$, is tradable.

Tradables and the market price of risk

The market price of risk is best introduced through a slight modification of the simple Black–Scholes model. That model had stock price $S_t = S_0 \exp(\sigma W_t + \mu t)$, and SDE

$$dS_t = S_t\left(\sigma\, dW_t + (\mu + \tfrac{1}{2}\sigma^2)\, dt\right).$$

We will find it convenient, however, to define price processes by means of their SDEs, typically

$$dS_t = S_t\left(\sigma\, dW_t + \mu\, dt\right),$$

which has solution $S_t = S_0 \exp(\sigma W_t + (\mu - \tfrac{1}{2}\sigma^2)t)$. The only difference between these two approaches is the subtraction of $\frac{1}{2}\sigma^2$ from the drift, which can be thought of as just a change of notation. Both forms can be equally used to define such geometric Brownian motions, but the SDE formulation allows a greater general class of models to be more easily considered.

Suppose then that we have a couple of tradable risky securities S_t^1 and S_t^2, both in the same market – that is both are functions of the same Brownian motion W_t, and both are defined via their SDEs,

$$dS_t^i = S_t^i(\sigma_i\, dW_t + \mu_i\, dt), \qquad i = 1, 2.$$

Following the discussion on tradables, we want the discounted prices of S_t^1 and S_t^2 to be martingales under the *same* measure $\mathbb{Q}$. So assuming a simple numeraire $B_t = \exp(rt)$, we have that

$$\tilde{W}_t = W_t + \left(\frac{\mu_i - r}{\sigma_i}\right) t$$

must be a $\mathbb{Q}$-Brownian motion for i equal to 1 and 2. But this can only happen if the two changes of drift are the same. That is if

$$\frac{\mu_1 - r}{\sigma_1} = \frac{\mu_2 - r}{\sigma_2}.$$

In one of those coincidences that cause confusion, economists attach a meaning to this quantity – if we interpret μ as the growth rate of the tradable, r as the growth rate of the riskless bond and σ as a measure of the risk of the asset, then

$$\gamma = \frac{\mu - r}{\sigma}$$

is the rate of extra return (above the risk-free rate) per unit of risk. As such it is often called the *market price of risk*.

Using this language then gives us a simple and compelling categorisation of tradables in terms of their SDEs – *all tradables in a market should have the same market price of risk.*

The general market price of risk

We can, in fact, generalise to more sophisticated one-factor models. Rigour will have to wait until section 6.1, but for now we can observe that a general stochastic price process S_t will have SDE

$$dS_t = S_t(\sigma_t \, dW_t + \mu_t \, dt),$$

where σ_t and μ_t are previsible processes.

Then defining

$$\gamma_t = \frac{\mu_t - r}{\sigma_t}$$

gives a time and state dependent market price of risk. Despite this variation, the same as above will hold. All tradable securities must instantaneously have the same market price of risk.

The risk-neutral measure

It is worth reflecting on what we have done – we have provided justification that to be tradable in a market defined by a stock S_t and a numeraire B_t is to share, after discounting by B_t, a martingale measure with S_t. This translates naturally in SDE terms to sharing a market price of risk – the market price of risk is actually the drift change of the underlying Brownian motion given by Cameron–Martin–Girsanov. So we have a natural means for sorting through SDEs for tradables.

We also have a natural explanation for the market terminology of $\mathbb{Q}$ as the *risk-neutral measure*. If we write the SDEs in terms of the $\mathbb{Q}$-Brownian motion $\tilde{W}_t$:

$$dS_t = S_t(\sigma_t \, d\tilde{W}_t + \tilde{\mu}_t \, dt),$$

then S_t is tradable if and only if its market price of risk is zero. All tradables then have the same growth rate under $\mathbb{Q}$ as the cash bond, independent of their riskiness σ_t – the measure $\mathbb{Q}$ is neutral with respect to risk.

But we should not overstretch the economic analogy – within our one-factor market all tradables are instantaneously perfectly correlated. They share a market price of risk not for profound economic reasons or because investors behave with certain risk preferences but for the reason that to do otherwise would produce a non-martingale process with a consequent opportunity for arbitrage. The market price of risk is only a convenient algebraic form for the change of measure from $\mathbb{P}$ to $\mathbb{Q}$, not a new argument for using it.

Non-tradable quantities

But convenient it is. Let's return to our underlying theme – dealing with non-tradable processes. With foreign exchange, equities and bonds we had a model for a process that had a fixed relationship to a tradable but was itself non-tradable. Concretely, we might have a non-tradable X_t which is modelled with the stochastic differential

$$dX_t = \sigma_t \, dW_t + \mu_t \, dt,$$

where σ_t and μ_t are previsible processes and W_t is $\mathbb{P}$-Brownian motion. Here σ_t and μ_t might be constants or constant multiples of X_t, but they needn't be.

We have X_t non-tradable but a deterministic function of X_t, $Y_t = f(X_t)$, is tradable. Then by Itô's formula, Y has differential increment

$$dY_t = \sigma_t f'(X_t) \, dW_t + \left(\mu_t f'(X_t) + \tfrac{1}{2}\sigma_t^2 f''(X_t)\right) dt.$$

4.4 Market price of risk

As Y_t is tradable, we can write down the market price of risk for Y_t immediately. Assuming the discount rate is constant at r,

$$\gamma_t = \frac{\mu_t f'(X_t) + \frac{1}{2}\sigma_t^2 f''(X_t) - rf(X_t)}{\sigma_t f'(X_t)}.$$

Since this market price of risk is simply the change of measure from $\mathbb{P}$ to $\mathbb{Q}$, we can write down X's behaviour under $\mathbb{Q}$ as

$$dX_t = \sigma_t \, d\tilde{W}_t + \frac{rf(X_t) - \frac{1}{2}\sigma_t^2 f''(X_t)}{f'(X_t)} \, dt.$$

Thus if we have claims on X_t, they can be priced via the normal expectation route, using this risk-neutral SDE for X_t.

Examples

(i) If X_t is the logarithm of a tradable asset, then f is the exponential function $f(x) = e^x$. In the simple case where $\sigma_t = \sigma$ and $\mu_t = \mu$ are constants (the basic Black–Scholes model), then the market price of risk for tradables is

$$\gamma_t = \frac{\mu + \frac{1}{2}\sigma^2 - r}{\sigma},$$

and the corresponding risk-neutral SDE for X_t is

$$dX_t = \sigma \, d\tilde{W}_t + (r - \frac{1}{2}\sigma^2) \, dt.$$

> **Time-dependent transforms**
>
> More generally, suppose interest rates follow the process r_t, X_t is non-tradable with stochastic differential
>
> $$dX_t = \sigma_t \, dW_t + \mu_t \, dt,$$
>
> and Y is a tradable security which is a deterministic function of X and time, that is $Y_t = f(X_t, t)$. Then under the martingale measure $\mathbb{Q}$, X has differential
>
> $$dX_t = \sigma_t \, d\tilde{W}_t + \frac{r_t f(X_t, t) - \frac{1}{2}\sigma_t^2 f''(X_t, t) - \partial_t f(X_t, t)}{f'(X_t, t)} \, dt,$$
>
> where f' and f'' are derivatives of f with respect to x, and $\partial_t f$ is the derivative of f with respect to t.

(ii)   The price process S_t pays dividends at rate δS_t. Let X_t be the process S_t and assume that it follows the Black–Scholes model

$$dX_t = X_t(\sigma\, dW_t + \mu\, dt).$$

The asset $Y_t = \exp(\delta t) X_t$ made from instantaneously reinvesting the dividends back into the stock holding is a tradable asset. The function f is thus chosen to be $f(x,t) = xe^{\delta t}$. The market price of risk for tradables is then

$$\gamma_t = \frac{\mu X_t e^{\delta t} + \delta X_t e^{\delta t} - r X_t e^{\delta t}}{\sigma X_t e^{\delta t}} = \frac{\mu + \delta - r}{\sigma},$$

and thus the risk-neutral SDE for X_t becomes

$$dX_t = X_t\left(\sigma\, d\tilde{W}_t + (r - \delta)\, dt\right).$$

(iii)   Foreign exchange, the 'wrong way round'. Let C_t be the dollar/mark exchange rate (worth in deutschmarks of one dollar), then the rate C_t *paid in dollars* is non-tradable. (That is, if C_t is equal to DM 1.45, the process worth \$1.45 is not tradable.) However the process $1/C_t$ is tradable, or more strictly e^{ut}/C_t is a dollar tradable asset if German interest rates are constant at rate u. If $X_t = C_t$ has SDE

$$dX_t = X_t(\sigma_t\, dW_t + \mu_t\, dt),$$

then the time-dependent transform of $f(x,t) = e^{ut}/x$ tells us that its risk-neutral SDE is

$$dX_t = X_t\left(\sigma_t\, d\tilde{W}_t + (\sigma_t^2 + u - r)\, dt\right).$$

4.5   Quantos

British Petroleum, a UK company, has a sterling denominated stock price. But instead of thinking of that stock price just in pounds, we could also consider it as a pure number which could be denominated in any currency.

4.5 Quantos

Contracts like this which pay off in the 'wrong' currency are *quantos*. For instance, if the current stock price were £5.20, we could have a derivative that paid this price in dollars, that is $5.20. This is not the same as the worth of the BP stock in dollars – that would depend on the exchange rate. What we have done is a purely formal change of units, whilst leaving the actual number unaltered.

Quantos are best described with examples. Here are three:

- a forward contract, namely receiving the BP stock price at time T as if it were in dollars in exchange for paying a pre-agreed dollar amount;

- a digital contract which pays one dollar at time T if the then BP stock price is larger than some pre-agreed strike;

- an option to receive the BP stock price less a strike price, in dollars.

In each case, a simple derivative is given the added twist of paying off in a currency other than in which the underlying security is denominated. And our intuition should warn us that this act of switching currency is not a foreign exchange quibble but something more fundamental. The British Petroleum stock price in dollars is a meaningful concept, but it is not a traded security. The payoffs we describe involve a non-tradable quantity.

Suppose we have a simple two-factor model. We have not actually met multi-factor models yet, but they are no more problematic than single-factor ones if we keep our head. Rigour can be found in the multiple stock models section (6.3). Our two random processes will be the stock price and the exchange rate, which will be driven by two independent Brownian motions $W_1(t)$ and $W_2(t)$.

For the construction, it is helpful to recall exercise 3.2: for ρ lying between -1 and 1, then $\rho W_1(t) + \sqrt{1 - \rho^2} W_2(t)$ is also a Brownian motion, and it has correlation ρ with the original Brownian motion $W_1(t)$. This is a useful way to manufacture two Brownian motions which are correlated out of a pair which are independent.

We suppose there exist the following constants: drifts μ and ν, positive volatilities σ_1 and σ_2, and a correlation ρ lying between -1 and 1.

Given these constants, the quanto model is:

Quanto model

The sterling stock price S_t and the value of one pound in dollars C_t follow the processes

$$S_t = S_0 \exp(\sigma_1 W_1(t) + \mu t),$$
$$C_t = C_0 \exp(\rho\sigma_2 W_1(t) + \bar{\rho}\sigma_2 W_2(t) + \nu t),$$

where $\bar{\rho}$ is the orthogonal complement of ρ, namely $\bar{\rho} = \sqrt{1 - \rho^2}$.

In addition there is a dollar cash bond $B_t = \exp(rt)$ and a sterling cash bond $D_t = \exp(ut)$, for some positive constant interest rates r and u.

Before we tease out the tradable instruments in dollars, note the co-variance of S_t and C_t. If we write our model in vector form, the vector random variable $(\log S_t, \log C_t)$ is jointly-normally distributed with mean vector $(\log S_0 + \mu t, \log C_0 + \nu t)$ and covariance matrix

$$\begin{pmatrix} \sigma_1 & 0 \\ \rho\sigma_2 & \bar{\rho}\sigma_2 \end{pmatrix} \begin{pmatrix} t & 0 \\ 0 & t \end{pmatrix} \begin{pmatrix} \sigma_1 & 0 \\ \rho\sigma_2 & \bar{\rho}\sigma_2 \end{pmatrix}^{\mathsf{T}} = \begin{pmatrix} \sigma_1^2 & \rho\sigma_1\sigma_2 \\ \rho\sigma_1\sigma_2 & \sigma_2^2 \end{pmatrix} t.$$

That is, we have ensured a constant volatility for S_t of σ_1, a constant volatility for C_t of σ_2 and a correlation between them of ρ.

Tradables

What are the dollar tradables? Following the intuition of the foreign exchange section (4.1), there are three: the dollar worth of the sterling bond, $C_t D_t$; the dollar worth of the stock, $C_t S_t$; and a dollar numeraire, the dollar cash bond B_t.

Writing down the first two of these tradables after discounting by the third, the numeraire, we have $Y_t = B_t^{-1} C_t D_t$ and $Z_t = B_t^{-1} C_t S_t$ respectively. Their SDEs are

$$dY_t = Y_t\left(\rho\sigma_2\, dW_1(t) + \bar{\rho}\sigma_2\, dW_2(t) + (\nu + \tfrac{1}{2}\sigma_2^2 + u - r)\, dt\right),$$
$$dZ_t = Z_t\big((\sigma_1 + \rho\sigma_2)\, dW_1(t) + \bar{\rho}\sigma_2\, dW_2(t)$$
$$+ (\mu + \nu + \tfrac{1}{2}\sigma_1^2 + \rho\sigma_1\sigma_2 + \tfrac{1}{2}\sigma_2^2 - r)\, dt\big).$$

(This can be checked using the n-factor Itô's formula of section 6.3.)

4.5 Quantos

As in the market price of risk section, we know we want to find a change of measure to make these martingales, or equivalently a market price of risk that represents this change of drift. As there are two sources of risk, $W_1(t)$ and $W_2(t)$, there will be two separate prices of risk. Respectively, $\gamma_1(t)$ will be the price of $W_1(t)$-risk and $\gamma_2(t)$ will be the price of $W_2(t)$-risk. In other words the market price of risk will be a vector $(\gamma_1(t), \gamma_2(t))$. We want to choose these γ so that the drift terms in dY_t and dZ_t vanish simultaneously. Not surprisingly this means solving a pair of simultaneous equations, or equivalently performing the matrix inversion

$$\begin{pmatrix} \gamma_1(t) \\ \gamma_2(t) \end{pmatrix} = \begin{pmatrix} \rho\sigma_2 & \bar{\rho}\sigma_2 \\ \sigma_1 + \rho\sigma_2 & \bar{\rho}\sigma_2 \end{pmatrix}^{-1} \begin{pmatrix} \nu + \frac{1}{2}\sigma_2^2 + u - r \\ \mu + \nu + \frac{1}{2}\sigma_1^2 + \rho\sigma_1\sigma_2 + \frac{1}{2}\sigma_2^2 - r \end{pmatrix}.$$

This is a particular case of the more general result that the multi-dimensional market price of risk is

$$\gamma_t = \Sigma^{-1}(\mu - r\mathbf{1}),$$

where Σ is the assets' volatility matrix, μ is their drift vector, and $\mathbf{1}$ is the constant vector $(1, \ldots, 1)$. More details are in section 6.3.

Here then we have a market price of risk $\gamma_t = (\gamma_1(t), \gamma_2(t))$, given by

$$\gamma_1 = \frac{\mu + \frac{1}{2}\sigma_1^2 + \rho\sigma_1\sigma_2 - u}{\sigma_1}, \quad \text{and} \quad \gamma_2 = \frac{\nu + \frac{1}{2}\sigma_2^2 + u - r - \rho\sigma_2\gamma_1}{\bar{\rho}\sigma_2}.$$

Thus under $\mathbb{Q}$ we can write the original processes S_t and C_t as

$$S_t = S_0 \exp\left(\sigma_1 \tilde{W}_1(t) + (u - \rho\sigma_1\sigma_2 - \tfrac{1}{2}\sigma_1^2)t\right),$$
$$C_t = C_0 \exp\left(\rho\sigma_2 \tilde{W}_1(t) + \bar{\rho}\sigma_2 \tilde{W}_2(t) + (r - u - \tfrac{1}{2}\sigma_2^2)t\right).$$

 Exercise 4.2 Verify that the measure $\mathbb{Q}$ which has Brownian motions $\tilde{W}_i(t) = W_i(t) + \int_0^t \gamma_i(s)\, ds$ $(i = 1, 2)$ really is the martingale measure for Y_t and Z_t.

Reassuringly the exchange rate process is as it was in section 4.1, given that $\rho\tilde{W}_1(t) + \bar{\rho}\tilde{W}_2(t)$ is another $\mathbb{Q}$-Brownian motion (as was proved in exercise 3.2).

But the stock price S_t is different from what we expected. The drift has an extra term: $-\rho\sigma_1\sigma_2$. For every value of ρ (except one, namely $(u-r)/\sigma_1\sigma_2$) this stops the dollar-discounted stock price being a $\mathbb{Q}$-martingale and thus prevents the price in dollars from being tradable. And that's precisely what our intuition warned us. There isn't a portfolio which is always worth a dollar amount numerically equal to the BP stock price.

Pricing

Since we have a measure $\mathbb{Q}$, under which the dollar tradables are martingales, we can price up our quanto options.

Forward

To price the forward contract, it helps to re-express the stock price at date T as

$$S_T = \exp(-\rho\sigma_1\sigma_2 T)F\exp(\sigma_1\sqrt{T}Z - \tfrac{1}{2}\sigma_1^2 T),$$

where F is the local currency forward price of S_T, $F = S_0 e^{uT}$, and Z is a normal $N(0, 1)$ random variable under $\mathbb{Q}$.

Then the value of the forward at time zero in dollars is

$$V_0 = e^{-rT}\mathbb{E}_{\mathbb{Q}}(S_T - k) = e^{-rT}\big(\exp(-\rho\sigma_1\sigma_2 T)F - k\big).$$

For this to be on market, that is to have a value of zero, we must set k to be $F\exp(-\rho\sigma_1\sigma_2 T)$. This is not the same as the simple forward price F for sterling purchase. As σ_1 and σ_2 are both positive, it is clear that this quanto forward price is greater than the simple forward price if and only if the correlation between the stock and the exchange rate is negative.

This actually makes some sense. Suppose we assumed that the quanto forward price was actually the same as the simple forward price F, then we could construct the following portfolio at time zero: by going

- long $C_0\exp\big((r-u)T\big)$ units of the quanto forward struck at F,

- short one unit of the simple sterling forward also struck at F.

If our assumption about the quanto forward price also being F were correct then this portfolio would be costless at time zero. At time T, this static replicating strategy would yield (in dollars)

$$C_0\exp\big((r-u)T\big)(S_T - F) - C_T(S_T - F) = (C_0\exp\big((r-u)T\big) - C_T)(S_T - F).$$

Noting that $C_0 \exp\big((r - u)T\big)$ is the forward FX rate for C_T, consider the effect of negative correlation. If the stock price ends up above its forward and the FX rate is below its forward, then the value of this portfolio is positive. And if the stock price ends up below F and the FX rate is above its forward, then the value is also positive.

Negative correlation makes these win–win situations more likely – perfect negative correlation makes them inevitable. If the quanto forward price really were F under these circumstances it wouldn't be hard to construct an arbitrage. For negative ρ the quanto forward must be greater than F.

Digital

Our digital contract, $I(S_T > k)$ in dollars, has price $V_0 = e^{-rT}\mathbb{Q}(S_T > k)$, or if we write $F_Q = F \exp(-\rho\sigma_1\sigma_2 T)$ the quanto forward price, then

$$V_0 = e^{-rT}\,\Phi\left(\frac{\log\frac{F_Q}{k} - \tfrac{1}{2}\sigma_1^2 T}{\sigma_1\sqrt{T}}\right).$$

Again the surprise of the $\exp(-\rho\sigma_1\sigma_2 T)$ term. And in a 'cleaner' option. Surely the event of S_T being greater than k is independent of whether the option is denominated in sterling or dollars. Indeed it is, but again replicating strategies, not expectation under $\mathbb{P}$, price options. And replication involves the exchange rate, which is correlated with the stock price.

Call option

Finally, we can compute the option price of $e^{-rT}\mathbb{E}_\mathbb{Q}\big((S_T - k)^+\big)$ as

$$V_0 = e^{-rT}\left(F_Q\Phi\left(\frac{\log\frac{F_Q}{k} + \tfrac{1}{2}\sigma_1^2 T}{\sigma_1\sqrt{T}}\right) - k\Phi\left(\frac{\log\frac{F_Q}{k} - \tfrac{1}{2}\sigma_1^2 T}{\sigma_1\sqrt{T}}\right)\right).$$

Perhaps not surprisingly for a log-normal model, this is just the original Black–Scholes formula with the quanto forward.

Exercise 4.3 Suppose everything remains the same, except that the stock S_t is the price in yen of NTT, a Japanese stock, C_t is the dollar/yen exchange rate (the worth in yen of one dollar), and ρ is their correlation. What is the one difference, between the sterling and yen cases, in the expression for the quanto forward price?

Chapter 5
Interest rates

Time is money. A dollar today is better than a dollar tomorrow. And a dollar tomorrow is better than a dollar next year. But just how much is that time worth – is every day worth the same or will the price of money change from time to time?

The interest rate market is where the price of money is set – how much does it cost to have money tomorrow, money in a year, money in ten years? Previously we made the modelling assumption that the cost of money is constant, but this isn't actually so. The price of money over a term depends not only on the length of the term, but also on the moment-to-moment random fluctuations of the interest rate market. In this way, money behaves just like a stock with a noisy price driven by a Brownian motion.

The uncertainty of the market opens up the possibility of derivative instruments based around the future value of money. Bonds, options on bonds, interest rate swaps, exotic contracts on the time value of different currencies, are all derived from basic interest-rate securities, just as stock options are derived from stocks in the market. In nominal cash terms, the market for such interest-rate derivatives far outstrips that for stock market products. Fortunately we shall still be able to calculate the prices of these contracts on exactly the same risk-free hedging basis as before.

5.1 The interest rate market

The most basic interest rate contract is an agreement to pay some money now in exchange for a promise of receiving a (usually) larger sum later. In general, the worth of such a contract will depend on factors other than just the time value of money, such as the credibility of the promisor and the perceived legality of the promise. Matters such as creditworthiness and the like are not our concern here, and it is for the bond market, not the interest rate market, to price them. We are solely concerned with the time value of money for default-free borrowing.

This basic contract only requires two numbers to describe it – its length, or *maturity*, which records when we are to receive the later payment, and the ratio of the size of that payment to our initial payment. We can call the maturity date on which we are paid T, and the fraction of the final payment which is the initial, $P(0, T)$. In other words, one dollar at time T can be bought at time zero for $P(0, T)$ dollars.

Discount bonds

But we can also regard the promise of a dollar as an asset, which will have some worth at time t before T. This asset is called a *discount bond*, and the price $P(0, T)$ is its price at time zero. But it can have a different price at any other time t up to maturity T, call it, say, $P(t, T)$. This price $P(t, T)$, the value at time t of receiving a dollar at time T, is a process in time – the price process of a tradable security.

For any one maturity T, the situation is much like the stock market in that here we have a tradable asset which has a stochastic price process. We feel we should be able to model its behaviour, and to price options on this T-bond by trading in it to hedge them. (The only difference is the technical point that the bond evolves towards a known value – at time T the bond is worth exactly one dollar, that is $P(T, T) = 1$. Stocks don't do things like that, but it won't turn out to be a problem.)

But we haven't got just one maturity. We could have written the contract for any one of the unlimited number of possible maturity dates. This matters because the bonds, although different, will be correlated. The ten-year bond, say, and the nine-year bond are going to move in very similar ways in the short term. Each bond cannot just be treated in isolation as if it were a stock. This is the real challenge of the interest rate market: the basic discount bonds are parameterised by two time indices, which determine both the start of the contract and its end. Bond prices are thus a function of two time variables, rather than just one, as stocks were.

The bond price graph is actually a two-dimensional surface lying in three-dimensional space, which we can explore by taking two-dimensional graphical sections through it. Illustrated are sections along the lines $t = 0$ (figure 5.1) and $T = 10$ (figure 5.2).

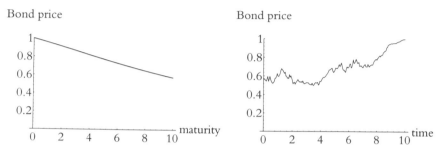

Fig. 5.1 Bond prices now   **Fig. 5.2** Price of 10-year bond vs time

Figure 5.1 is not the price process of an asset, but a graph of the current price of a whole spectrum of different assets (the bonds of different maturities). This reflects the current time value of money, quantifying exactly how much better it is to have cash now rather than later. Generally the more distant the payment maturity date, the less the current worth of the bond. Figure 5.2 is the price of one particular asset (the ten-year discount bond). Now instead of a smooth graph, we have a noisy stochastic process, up until it hits the value one at its maturity time. The start point of this graph is the end point of figure 5.1, being the common value $P(0, 10)$, or the worth now of receiving a dollar in ten years' time.

Yields

The picture in figure 5.1 is not particularly sensitive to what the market is doing. Other than saying now is better than later, it doesn't tell us very much on quick inspection. A more informative measure of the market is an indication of the implied average interest rate offered by a bond. If interest rates were constant at rate r, the price of the T-bond at time t would be $e^{-r(T-t)}$. In this particular case, r can be recovered from the price $P(t, T)$ via the formula $r = -\log P(t, T)/(T - t)$.

Interest rates are not constant, but that doesn't stop us viewing this translation as potentially useful. The rate we derive, $R(t,T)$, is called the *yield*, and the mapping from price to yield is one-to-one for t less than T — no information is lost.

Yield

Given a discount bond price $P(t,T)$ at time t, the yield $R(t,T)$ is given by

$$R(t,T) = -\frac{\log P(t,T)}{T-t}.$$

Thus for any given discount bond price curve, we can produce a *yield curve*; that is, a graph of $R(t,T)$ against T for some fixed t.

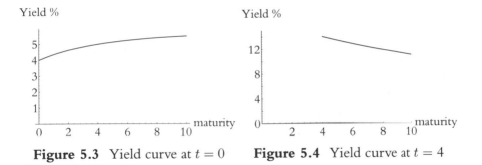

Figure 5.3 Yield curve at $t = 0$ **Figure 5.4** Yield curve at $t = 4$

While the discount bond price curve contains exactly the same information as the yield curve, the translation is friendlier to the eye. Long dated bonds always have lower prices, so the downwards slope of the price curve is inevitable, thus redundant. Yield curves, on the other hand, can be increasing or decreasing functions of T, revealing the average return of bonds stripped of the crude effects of maturity — the *term structure* of the market.

The difference in yields at different maturities reflects market beliefs about future interest rates. If there is a possibility that rates might be higher in the future, long-term loans will have to charge a higher rate than short-term ones. Typically the yield does increase with maturity, due to increased uncertainty about far-distant interest rates, but if current rates are high and

131

expected to fall, the yield curve can become 'inverted' and long bond yields will be less than short bonds (figure 5.4). A good model should be able to cope with both these possibilities.

Instantaneous rate

But what is the price of money now? The yield curve gives us an idea of the rate of borrowing for each term length, but it would be convenient if we could summarise the *current* cost of borrowing in a single number. What we can do is look at the current rate for instantaneous borrowing. That is, borrowing which is paid back (nearly) instantly. If at time t we borrow over the period from t to $t + \Delta t$, where Δt is a small time increment, the rate we get is the yield $R(t, t + \Delta t)$:

$$R(t, t + \Delta t) = -\frac{\log P(t, t + \Delta t)}{\Delta t}.$$

For ever smaller time increments this value more closely approximates to $R(t, t)$, which is the left-most point of the yield curve at time t. We call this value the *instantaneous rate*, or *short rate*, r_t, which is given by both the expressions

$$r_t = R(t, t),$$

$$\text{and} \quad r_t = -\frac{\partial}{\partial T} \log P(t, t).$$

The instantaneous rate is just a process in time, free of any other parameters. Figure 5.5 shows an example short rate over ten years, corresponding to the evolution of the 10-year bond in figure 5.2.

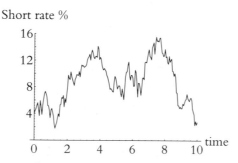

Figure 5.5 Instantaneous rate

We can sometimes see an interaction between the short rate and the bond prices if they are correlated. In one instance, bond prices might be lower when the short rate is higher, which can be seen in this example around the 4 year and 8 year marks, when the short rate gets high and the bond price dips. Interestingly though, the high short rate at $t = 4$ even exceeds the increased yields on longer bonds, giving an inverted curve (figure 5.4).

The instantaneous rate is not only an important process in the interest rate market, but many models are based exclusively on its behaviour, with all the other bonds extrapolated from it.

Forwards

The short-rate process, r_t, is not a one-to-one mapping from the discount price curve $P(t, T)$. The translation also entails a loss of information. Just giving r_t with no extra prescription on how bond prices can move will not in general be enough to recover $P(t, T)$. Yet the instantaneous rate is convenient to work with. What we require is a natural extension of r_t which brings back the one-to-one mapping to the prices $P(t, T)$ and the yields $R(t, T)$, yet still preserves the idea of instantaneity.

Consider forward contracts, that is agreeing, at time t, to make a payment at a later date T_1 and receive a payment in return at an even later date T_2. We are really just striking a forward on the T_2-bond. But what forward price should we pay?

There is a way of replicating this contract by, at time t, buying a T_2-bond and selling a quantity, say k units, of the T_1-bond. This deal has initial cost $P(t, T_2) - kP(t, T_1)$ at time t, and will require us to make a payment of k at time T_1, and will give us a payment of one dollar at time T_2. To give the contract nil initial value, we must set k to be

$$k = \frac{P(t, T_2)}{P(t, T_1)}.$$

This k must be, or face arbitrage, the forward price of purchasing the T_2-bond at time T_1. The corresponding (forward) yield is then

$$-\frac{\log P(t, T_2) - \log P(t, T_1)}{T_2 - T_1}.$$

Were we to choose T_1 and T_2 very close together, say $T_1 = T$ and $T_2 = T + \Delta t$, then as the increment Δt became smaller this would converge to a forward rate for instantaneous borrowing,

$$f(t, T) = -\frac{\partial}{\partial T} \log P(t, T).$$

This rate, called simply the *forward rate*, is the forward price of instantaneous borrowing at time T. As we might expect, the 'forward' rate for borrowing now, at time $T = t$, is exactly the current instantaneous rate, that is

$$f(t, t) = r_t.$$

But unlike r_t, given the forward rates $f(t, T)$ we can recover the prices $P(t, T)$ and the yields $R(t, T)$. The translation $f(t, T)$ for our particular example is shown in figure 5.6.

Superficially, the forward rate curve resembles the bond yield curve (figure 5.3). Indeed the yield curve and the forward curve agree at their leftmost point, the instantaneous rate, but the other points of the two curves will generally be different. But the formula for $R(t, T)$ can be differentiated and rearranged to show that

$$f(t, T) = R(t, T) + (T - t)\frac{\partial R}{\partial T}(t, T).$$

This tells us that the forward rate curve is higher than the yield curve, if the yield curve is increasing, and lower than it if the yield curve is inverted.

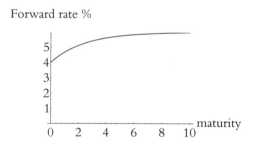

Figure 5.6 Forward rate curve at time $t = 0$

As a function of time rather than maturity the forward rate will not be so smooth, but will start with some initial value $f(0, T)$ and evolve as a stochastic process, finishing with the value r_T at time T.

Summary

We have a market of default-free zero coupon discount bonds. The price at time t of the T-bond which pays off one dollar at time T is $P(t,T)$. The average yield of the bond over its remaining lifetime is $R(t,T)$, and the price now of instantaneous borrowing at time T is the forward rate $f(t,T)$. The price of instantaneous spot borrowing is $r_t = R(t,t) = f(t,t)$.

Both of these associated families of rates, $R(t,T)$ and $f(t,T)$, contain all the original price information which can be recovered. Explicitly

Interest-rate market summary

The forward rates and the yield can be written in terms of the bond prices as

$$f(t,T) = -\frac{\partial}{\partial T} \log P(t,T), \quad \text{and} \quad R(t,T) = -\frac{\log P(t,T)}{T-t}.$$

And conversely, the bond prices can be given in terms of the forward rates or the yields:

$$P(t,T) = \exp\left(-\int_t^T f(t,u)\,du\right),$$

and

$$P(t,T) = \exp\left(-(T-t)R(t,T)\right).$$

In other words, for modelling purposes we can choose to specify the behaviour of only any one of these three, and the other two will follow automatically.

5.2 A simple model

A concrete example is illuminating. The secret of this chapter is that we can tackle interest rate models in exactly the same way as stock models. The Itô manipulations are harder but they are not significant for the story — just as in Black–Scholes, the real work is carried by the martingale representation theorem. In Black–Scholes, there were only two canonical tradables (the stock and the bond), but there are now an infinite number of underlying discount bonds. To pick just two of these tradables to work with would seem to favour that pair over the rest, but such worries will prove illusory. All the tradables will still turn out to be martingales under the risk-neutral measure, which itself is independent of the apparent 'choice' of instruments to work with.

Simple interest rate model

Given an initial T-integrable forward rate curve $f(0,T)$, the forward curve evolves as:

$$d_t f(t,T) = \sigma\, dW_t + \alpha(t,T)\, dt,$$

for some constant volatility σ and drift α, a bounded deterministic function of time and maturity.

We have set the market up not with an SDE for the price of any asset, but with the SDE for the forward rate. However as chapter four has shown, as long as we are an Itô step away from the price of something, this doesn't have to pose a problem.

The forward rate itself is

$$f(t,T) = \sigma W_t + f(0,T) + \int_0^t \alpha(s,T)\, ds.$$

Thus the forward rate is normally distributed. Moreover the forward rates at different maturities are perfectly correlated in their movements as the difference between any two of them, $f(t,T) - f(t,S)$, is purely deterministic. There is only one source of randomness, the Brownian motion W_t, and that is a process over time, not over maturity.

Tradable securities

Tradables may only be an Itô step away, but what are they? One is obvious – we want a numeraire. Though chapter six will show the choice of numeraire doesn't really matter, there is a canonical candidate – the cash bond formed by the instantaneous rate r_t. That is, B_t given by

$$dB_t = r_t B_t\, dt, \qquad B_0 = 1.$$

Since r_t is given by $f(t,t)$, we can write down its integral equation easily enough as

$$r_t = \sigma W_t + f(0,t) + \int_0^t \alpha(s,t)\, ds.$$

[Technical trap: the SDE for r_t is not just the SDE for $f(t,T)$ 'evaluated' at $T = t$. Itô tells us it is actually $dr_t = d_t f(t,t) + \frac{\partial f}{\partial T}(t,t)$.] Unlike the basic stock models, this rate is not constant but rather is a random process, and it is normally distributed, which admits the possibility of it being negative sometimes. Later we will show models which overcome this, but for the moment we'll pay this price for simplicity. Now for the cash bond, $B_t = \exp(\int_0^t r_s\, ds)$, which has the slightly daunting expression

$$B_t = \exp\left(\sigma \int_0^t W_s\, ds + \int_0^t f(0,u)\, du + \int_0^t \int_s^t \alpha(s,u)\, du\, ds\right).$$

This will be our tradable numeraire.

What about another tradable? Here, as mentioned earlier, there's an embarrassment of tradables, but let's pick one. Fixing T, we have the price of the T-maturity bond $P(t,T)$ given by $P(t,T) = \exp(-\int_t^T f(t,u)\, du)$ or

$$P(t,T) = \exp-\left(\sigma(T-t)W_t + \int_t^T f(0,u)\, du + \int_0^t \int_t^T \alpha(s,u)\, du\, ds\right).$$

Replicating strategies

Suppose then we wanted to replicate some claim X at a time horizon S less than T (so that the T-maturity bond doesn't vanish on us). In chapters two, three and four we had a three-stage replicating strategy, so at least a first guess would be to follow it here as well:

Three steps to replication (interest-rate market)

(1) Find a measure $\mathbb{Q}$ under which the T-bond discounted by the cash bond $Z_t = B_t^{-1} P(t,T)$ is a martingale.

(2) Form the process $E_t = \mathbb{E}_{\mathbb{Q}}(B_S^{-1} X \mid \mathcal{F}_t)$.

(3) Find a previsible process ϕ_t, such that $dE_t = \phi_t\, dZ_t$.

First we tackle the complicated-looking discounted bond price process $Z_t = B_t^{-1} P(t, T)$:

$$Z_t = \exp -\left(\sigma(T-t)W_t + \sigma \int_0^t W_s \, ds + \int_0^T f(0, u) \, du + \int_0^t \int_s^T \alpha(s, u) \, du \, ds \right).$$

Noting that the $\sigma(T-t)W_t$ term's differential can be handled with the product rule and everything else inside the exponential is either constant or easy to differentiate, Itô gives us the SDE for Z_t as

$$dZ_t = Z_t \left(-\sigma(T-t) \, dW_t - \left(\int_t^T \alpha(t, u) \, du \right) dt + \tfrac{1}{2}\sigma^2(T-t)^2 \, dt \right).$$

Now we are on familiar ground. Though we are used to the cash bond B_t being deterministic, this (random) B_t and the T-bond price $P(t, T)$ are both adapted to the same Brownian motion, W_t, and finding Z_t doesn't pose any real problem.

Step one

We have an SDE for Z_t and we want to see if we can find a change of measure drift γ_t for the Brownian motion which makes Z_t driftless. The candidate is clearly

$$\gamma_t = -\tfrac{1}{2}\sigma(T-t) + \frac{1}{\sigma(T-t)} \int_t^T \alpha(t, u) \, du.$$

And since it is bounded up to time S, the technical conditions of C–M–G are satisfied – our candidate passes and we have a measure $\mathbb{Q}$, equivalent to $\mathbb{P}$, such that $\tilde{W}_t = W_t + \int_0^t \gamma_s \, ds$ is a $\mathbb{Q}$–Brownian motion. The SDE for the discounted price Z_t now becomes

$$dZ_t = -\sigma Z_t(T-t) \, d\tilde{W}_t.$$

The process Z_t has no drift, and because $\sigma(T-t)$ is bounded up to time S, Z_t is a $\mathbb{Q}$-martingale.

Step two

This gives us E_t as the conditional $\mathbb{Q}$-expectation of the discounted claim $B_S^{-1}X$, namely

$$E_t = \mathbb{E}_\mathbb{Q}\left(B_S^{-1}X \mid \mathcal{F}_t \right).$$

But since E_t is a $\mathbb{Q}$-martingale just as Z_t is, we take:

5.2 A simple model

Step three

Using the martingale representation theorem to link them via an $\mathcal{F}$-previsible process ϕ_t:

$$E_t = \mathbb{E}_{\mathbb{Q}}(B_S^{-1}X) + \int_0^t \phi_s \, dZ_s.$$

What is our trading strategy? At time t,

- hold ϕ_t units of the T-bond $P(t,T)$

- hold $\psi_t = E_t - \phi_t Z_t$ units of the cash bond B_t.

The undiscounted value of this portfolio at time t is

$$V_t = \phi_t P(t,T) + \psi_t B_t = B_t E_t.$$

As before, it is also true that $dV_t = \phi_t \, d_t P(t,T) + \psi_t \, dB_t$ and thus this portfolio is self financing. The strategy has an initial cost of $V_0 = \mathbb{E}_{\mathbb{Q}}(B_S^{-1}X)$ and has a terminal value $V_S = X$, which exactly hedges the claim. Arbitrage has won through.

> **Option price formula (interest rate)**
> The price of X at time t is
> $$V_t = B_t \, \mathbb{E}_{\mathbb{Q}}(B_S^{-1}X \mid \mathcal{F}_t).$$

No free lunches

So far, so good – even though the Itô work was harder, we have just another stock-type model. The chosen pair B_t and $P(t,T)$ behaved like any of the tradables of chapter four. But something *should* worry us. We picked a particular bond, the T-maturity bond, and found a change of measure particular to that. Yet all claims which paid off at time S before T could be hedged, even those, for example, which are identical to bond of *other* maturities.

So we have *two* ways of pricing the S-bond at time t, $P(t,S)$. One direct from its SDE. And the other indirect, viewing $X = P(S,S) = 1$ as a claim to be hedged via the cash bond and the T-bond.

There is no obvious reason why they should be the same given our original model. And yet the same they must be. If the hedge price were ever, say, less than $P(t, S)$ we would have an arbitrage engine capable of locking in unlimited profits. We don't want free lunches, so we should assume that the real world forbids them. That is we should impose on our real world model some suitable condition to make the various ways of getting at the price $P(t, S)$ agree. What condition?

Consider the discounted process of the S-bond, $Y_t = B_t^{-1} P(t, S)$. Re-working the Itô from before we have, as expected,

$$dY_t = Y_t \left(-\sigma(S - t)\, dW_t - \left(\int_t^S \alpha(t, u)\, du \right) dt + \tfrac{1}{2} \sigma^2 (S - t)^2\, dt \right).$$

If we define γ_t^S to be

$$\gamma_t^S = -\tfrac{1}{2} \sigma (S - t) + \frac{1}{\sigma(S - t)} \int_t^S \alpha(t, u)\, du,$$

then we have $dY_t = -\sigma Y_t (S - t) \big(dW_t + \gamma_t^S\, dt \big)$, or in terms of the $\mathbb{Q}$-Brownian motion we had before:

$$dY_t = -\sigma Y_t (S - t) \big(d\tilde{W}_t + (\gamma_t^S - \gamma_t)\, dt \big).$$

This discounted process Y_t *must* be a $\mathbb{Q}$-martingale – it's tradable and, from the risk-free hedging construction, $Y_t = B_t^{-1} P(t, S) = \mathbb{E}_{\mathbb{Q}}(B_S^{-1} | \mathcal{F}_t)$. So the drift term of the SDE above must be zero: $\gamma_t^S = \gamma_t$.

Here is the restriction we require – our arbitrary choice of T must not have affected the process γ_t. So γ_t must be independent of T, or in other words $\frac{\partial \gamma}{\partial T} = 0$.

Multiplying the formula for γ_t by $\sigma(T - t)$ and differentiating with respect to T, we get:

Restriction on the drift

In an arbitrage-free market, the drift $\alpha(t, T)$ satisfies

$$\alpha(t, T) = \sigma^2 (T - t) + \sigma \gamma_t.$$

This equation is saying something we did not encounter, at this level, in the stock market. It says that there are restrictions on the drifts which the forward rates can have if there is to be no arbitrage. The drift $\alpha(t,T)$ may have started off as a general deterministic function of both time and maturity, but now it is expressed as the sum of a particular function $(\sigma^2(T-t))$ and a process which has no maturity dependence at all $(\sigma\gamma_t)$. Most general functions cannot be written in this way.

In another sense, this is actually familiar ground. We can think of the SDE for $P(t,T)$ under $\mathbb{P}$ as

$$d_t P(t,T) = P(t,T)\Big(-\sigma(T-t)\,dW_t + \big(r_t - \sigma(T-t)\gamma_t\big)\,dt\Big).$$

Written this way, γ_t stands revealed as the market price of risk (see section 4.4). We know that every security in the market has to have the same market price of risk, which explains why γ_t does not depend on the maturity T chosen.

Two things stand out. Firstly there is a measure $\mathbb{Q}$ which makes a martingale not just out of one discounted bond, but each and every discounted bond simultaneously. We worried about the embarrassment of bonds to choose, but we needn't have. There was only one Brownian motion and that is what matters. If we freeze time and look at just one t, the values of the bonds $P(t,T)$ are just deterministic transformations of each other. And if one bond can be brought into line by a change of measure then so can they all.

If, that is, they are roughly in step in the first place. Our second point is that there is a price to pay for this success. If we write the original SDE for $f(t,T)$ in terms of the $\mathbb{Q}$-Brownian motion, $\tilde{W}_t$, we have:

$$d_t f(t,T) = \sigma\,d\tilde{W}_t + \sigma^2(T-t)\,dt.$$

As we expect from a Black–Scholes upbringing, the drift $\alpha(t,T)$ has vanished. But $\alpha(t,T)$ must be recoverable by a change of measure γ_t which has no dependence on T. So we weren't free to pick $\alpha(t,T)$ as any function of t and T – we must, unlike Black–Scholes, have some structure to the original real world drift.

But even if our success has brought slight complications, we have nonetheless succeeded. We have a model with stochastic interest rates which is still arbitrage-complete. All claims can be coherently hedged by the underlying bonds. Once more, replication provides the price.

Bonds and rates in terms of the $\mathbb{Q}$-Brownian motion $\tilde{W}_t$

The bond prices, forward and short rates are given by:

$$P(t,T) = \exp - \left(\sigma(T-t)\tilde{W}_t + \int_t^T f(0,u)\,du + \tfrac{1}{2}\sigma^2 T(T-t)t \right),$$

$$B_t = \exp \left(\sigma \int_0^t \tilde{W}_s\,ds + \int_0^t f(0,u)\,du + \tfrac{1}{6}\sigma^2 t^3 \right),$$

$$f(t,T) = \sigma\tilde{W}_t + f(0,T) + \sigma^2(T - \tfrac{1}{2}t)t,$$

$$r_t = \sigma\tilde{W}_t + f(0,t) + \tfrac{1}{2}\sigma^2 t^2.$$

5.3 Single-factor HJM

From the particular to the general. We know the basic idea – all three descriptions of the yield curve, the prices $P(t,T)$, the yields $R(t,T)$ and the forward rates $f(t,T)$ are equivalent, so we select one and specify its behaviour. Heath–Jarrow–Morton (HJM) is a powerful, technically rigorous interest-rate model based on the instantaneous forward rates $f(t,T)$.

Single-factor HJM model

Given an initial forward rate curve $f(0,T)$, the forward rate for each maturity T evolves as

$$f(t,T) = f(0,T) + \int_0^t \sigma(s,T)\,dW_s + \int_0^t \alpha(s,T)\,ds, \qquad 0 \leqslant t \leqslant T,$$

or in differential form

$$d_t f(t,T) = \sigma(t,T)\,dW_t + \alpha(t,T)\,dt.$$

The volatilities $\sigma(t,T)$ and the drifts $\alpha(t,T)$ can depend on the history of the Brownian motion W_t and on the rates themselves up to time t.

For any fixed maturity T, the forward rate evolves according to its own volatility $\sigma(t,T)$ and its own drift $\alpha(t,T)$. In section 5.5 we will allow the decoupling that comes when rates can move with less than perfect correlation, but here only a single process, a $\mathbb{P}$-Brownian motion W_t, will drive each and every rate. The incremental changes of all forward rates, and thus all yields and all bond prices are perfectly correlated.

Our formal description is vague about the precise properties of the volatility and drift functions. The general HJM model posits very few overarching conditions on the σ and α, but imposes piecemeal technical constraints from time to time. Collected, and simplified somewhat, these technical conditions are shown in the box. The first two conditions make sure that the forward rates $f(t,T)$ are well defined by their SDE. The last two conditions will be used for a Fubini-type result that the stochastic differential of the integral of $f(t,T)$ with respect to T is the integral of the stochastic differentials of f. Given these box conditions, the first three conditions of the HJM model (C1–C3 in their paper) are satisfied.

Single-factor HJM: conditions on the volatility and drift

We assume that

- for each T, the processes $\sigma(t,T)$ and $\alpha(t,T)$ are previsible and depend only on the history of the Brownian motion up to time t, and are good integrators in the sense that $\int_0^T \sigma^2(t,T)\,dt$ and $\int_0^T |\alpha(t,T)|\,dt$ are finite;

- the initial forward curve, $f(0,T)$, is deterministic and satisfies the condition that $\int_0^T |f(0,u)|\,du < \infty$;

- the drift α has finite integral $\int_0^T \int_0^u |\alpha(t,u)|\,dt\,du$;

- the volatility σ has finite expectation $\mathbb{E}\int_0^T \left|\int_0^u \sigma(t,u)\,dW_t\right|\,du$.

Numeraire

As chapter six will show, the choice of numeraire is arbitrary – but algebraic convenience certainly points to a canonical choice. Our description of the forward rates $f(t,T)$ allows us to write down an integral equation for the instantaneous rate $r_t = f(t, t)$ (which need not be Markov), namely:

$$r_t = f(0, t) + \int_0^t \sigma(s, t) \, dW_s + \int_0^t \alpha(s, t) \, ds.$$

The simplest cash product is then the account, or bond, formed by starting with one dollar at time zero and reinvesting continually at this rate. In other words, the bond B is a stochastic process satisfying the SDE

$$dB_t = r_t B_t \, dt, \quad B_0 = 1, \quad \text{or} \quad B_t = \exp\left(\int_0^t r_s \, ds\right).$$

Integration then gives us

$$B_t = \exp\left(\int_0^t \left(\int_s^t \sigma(s, u) \, du\right) dW_s + \int_0^t f(0, u) \, du + \int_0^t \int_s^t \alpha(s, u) \, du \, ds\right).$$

Here we used the last technical condition of the HJM box to say that the integrals of $\int_0^t \left(\int_0^u \sigma(s, u) \, dW_s\right) du$ can be interchanged to $\int_0^t \left(\int_s^t o(s, u) \, du\right) dW_s$. We have a numeraire.

Bond prices

We need tradable assets – and we have them, the bonds $P(t, T)$. Since the forward rates $f(t, T)$ are a one-to-one transformation, the bond prices themselves are contained in the forward rate information as

$$P(t, T) = \exp\left(-\int_t^T f(t, u) \, du\right),$$

which will be continuous in t and T. If we integrate the original equation for the forward rates $f(t, T)$ then we have the bond price $P(t, T)$ equal to

$$\exp - \left(\int_0^t \left(\int_t^T \sigma(s, u) \, du\right) dW_s + \int_t^T f(0, u) \, du + \int_0^t \int_t^T \alpha(s, u) \, du \, ds\right).$$

Reassuringly this expression, although awkward, has the right values at time zero (namely $\exp\left(-\int_0^T f(0, u) \, du\right)$) and time T (namely one).

Discounted bonds

Let's fix one particular maturity T to work with for the moment. As everywhere else, our attention focuses on the discounted asset price — that is, $Z(t,T) = B_t^{-1} P(t,T)$. By combining the above expressions for the cash bond and the bond price itself, we get

$$Z(t,T) = \exp \left(\int_0^t \Sigma(s,T)\,dW_s - \int_0^T f(0,u)\,du - \int_0^t \int_s^T \alpha(s,u)\,du\,ds \right),$$

where $\Sigma(t,T)$ is just notation for the integral $-\int_t^T \sigma(t,u)\,du$. Itô handle-turning then gives the SDE —

$$d_t Z(t,T) = Z(t,T) \left(\Sigma(t,T)\,dW_t + \left(\tfrac{1}{2}\Sigma^2(t,T) - \int_t^T \alpha(t,u)\,du \right) dt \right),$$

revealing the variable $\Sigma(t,T)$ to be the log-volatility of $P(t,T)$.

Change of measure

In the usual way, we want to make the discounted bond into a martingale by changing measure. The change of measure drift (market price of risk) is

$$\gamma_t = \tfrac{1}{2}\Sigma(t,T) - \frac{1}{\Sigma(t,T)} \int_t^T \alpha(t,u)\,du.$$

We need the technical Cameron–Martin–Girsanov theorem condition that $\mathbb{E}_{\mathbb{P}} \exp \tfrac{1}{2} \int_0^T \gamma_t^2\,dt$ is finite. Then there will be a new measure $\mathbb{Q}$ equivalent to $\mathbb{P}$, such that $\tilde{W}_t = W_t + \int_0^t \gamma_s\,ds$ is $\mathbb{Q}$-Brownian motion. The SDE of the discounted bond under $\mathbb{Q}$ is then

$$d_t Z(t,T) = Z(t,T)\Sigma(t,T)\,d\tilde{W}_t,$$

which is driftless. For this to be a proper $\mathbb{Q}$-martingale, it is sufficient that the exponential martingale condition $\mathbb{E}_{\mathbb{Q}} \exp \tfrac{1}{2} \int_0^T \Sigma^2(t,T)\,dt < \infty$ holds (see section 3.5).

> ### Bond price SDE
> Under this martingale measure, the bond price P now has the stochastic differential
>
> $$d_t P(t,T) = P(t,T) \left(\Sigma(t,T)\,d\tilde{W}_t + r_t\,dt \right).$$

The concrete model of section 5.2 partially spoilt the surprise, but we have our Black–Scholes like result, even here with a general interest-rate model such as HJM. The behaviour of the price P under the martingale measure does not depend on the drift α, but only on the volatility Σ (itself a function of σ). Just as the Black–Scholes stock model under $\mathbb{Q}$ had no dependence on the original stock drift μ.

Replicating strategies

We've jumped slightly ahead of ourselves, we have found the martingale measure and the process for $P(t,T)$ under it. But we ought to check that we can produce replicating strategies for claims. Suppose we have a claim X which pays off at time S. If we are going to hedge this with a discount bond maturing at date T, our only restriction is that S should come before T – we cannot hedge a long-term product with a shorter-term instrument. (Unless we split the time-period up into shorter subsections, and roll over short-term bonds from section to section.) Suppose, for simplicity, we choose to use a bond with maturity T larger than S.

As before, our second step to replication is to form the conditional $\mathbb{Q}$-expectation of the discounted claim $B_S^{-1}X$, rather than the raw claim X. That is, we define E_t to be the $\mathbb{Q}$-martingale

$$E_t = \mathbb{E}_{\mathbb{Q}}\big(B_S^{-1}X \mid \mathcal{F}_t\big).$$

For the martingale representation theorem to be used, we also need that the bond volatility $\Sigma(t,T)$ is never zero before T, in which case, we apply the representation theorem to the martingale $Z(t,T)$ and the discounted claim process E_t. This gives us that

$$E_t = E_0 + \int_0^t \phi_s \, dZ(s,T),$$

for some $\mathcal{F}$-previsible process ϕ.

Our trading strategy will be a combination of both a holding in the T-bond and a holding in the cash bond B_t. Specifically, we

- hold ϕ_t units of the T-bond at time t,

- hold $\psi_t := E_t - \phi_t Z(t,T)$ units of the cash bond at time t.

The value of this portfolio at time t is

$$V_t = B_t E_t = B_t \, \mathbb{E}_\mathbb{Q} \big(B_S^{-1} X \mid \mathcal{F}_t \big).$$

The strategy (ϕ_t, ψ_t) will be self-financing if $dV_t = \phi_t \, d_t P(t, T) + \psi_t \, dB_t$, or equivalently (as in section 3.7) if

$$dE_t = \phi_t \, d_t Z(t, T).$$

Which is ensured by the representation of E_t in terms of ϕ_t. The portfolio (ϕ_t, ψ_t) is self-financing. Thus:

Derivative pricing

If X is the payoff of a derivative maturing at time T, then its value at time t is

$$V_t = B_t \, \mathbb{E}_\mathbb{Q} \big(B_T^{-1} X \mid \mathcal{F}_t \big) = \mathbb{E}_\mathbb{Q} \Big(\exp\big(-\textstyle\int_t^T r_s \, ds\big) X \mid \mathcal{F}_t \Big).$$

Arbitrage-free market

But the S-bond is simply a claim of $X = 1$ maturing at time S. Thus its worth at time t must be $B_t \mathbb{E}_\mathbb{Q}(B_S^{-1}|\mathcal{F}_t)$. Or more fully,

$$P(t, S) = \mathbb{E}_\mathbb{Q} \left(\exp\Big(-\int_t^S r_u \, du\Big) \,\Big|\, \mathcal{F}_t \right), \qquad t \leqslant S < T.$$

The martingale measure brings a pleasant simplicity. All bond prices are just the expectation under $\mathbb{Q}$ of the instantaneous rate discount from t to their maturity.

What about the discounted S-bond, $Z(t, S) = B_t^{-1} P(t, S)$? This can now be written as

$$Z(t, S) = \mathbb{E}_\mathbb{Q} \big(B_S^{-1} \mid \mathcal{F}_t \big).$$

Just as we saw in the simple example (section 5.2), all the other (discounted) bonds are now martingales under the same $\mathbb{Q}$. Which means that their drifts under $\mathbb{P}$ are restricted by the need to be a simple change of measure away from a martingale. In other words, the market price of risk has to be the same for all bonds, or else there will be an arbitrage opportunity.

So we have a restriction on the bonds' $\mathbb{P}$-drifts. In particular, it must be the case that, for all maturities T,

$$\int_t^T \alpha(t,u)\,du = \tfrac{1}{2}\Sigma^2(t,T) - \Sigma(t,T)\gamma_t, \qquad t \leqslant T.$$

Differentiating with respect to T, we see that $\alpha(t,T) = -\sigma(t,T)\Sigma(t,T) + \sigma(t,T)\gamma_t$, that is

$$\alpha(t,T) = \sigma(t,T)\big(\gamma_t - \Sigma(t,T)\big).$$

Exactly as in section 5.2, where $\sigma(t,T) = \sigma$ and $\Sigma(t,T) = -\sigma(T-t)$, the real world drift $\alpha(t,T)$ cannot be too different from the risk–neutral value of $-\sigma(t,T)\Sigma(t,T)$.

Under this risk–neutral measure, the forward rate and the instantaneous rate are then,

> **Forward and short rates under $\mathbb{Q}$**
>
> $$d_t f(t,T) = \sigma(t,T)\,d\tilde{W}_t - \sigma(t,T)\Sigma(t,T)\,dt,$$
> $$r_t = f(0,t) + \int_0^t \sigma(s,t)\,d\tilde{W}_s - \int_0^t \sigma(s,t)\Sigma(s,t)\,ds.$$

Like the bond price itself, these expressions no longer depend on the drift at all, but are solely expressed in terms of the volatilities σ and Σ.

Model conditions

We have been accumulating technical conditions as we have swept through. They are summarised in the box below.

The first condition is actually necessary and sufficient for there to be an equivalent measure under which every single discounted bond price is a martingale, which guarantees the absence of arbitrage. The second condition is equivalent to asserting that the change of measure is unique, which means that all risks can be hedged using the martingale representation theorem. The last two conditions are technical requirements for C–M–G to operate and to make sure that Z is a martingale under the new measure.

Single-factor HJM: market completeness conditions

It is required that

- there exists an $\mathcal{F}$-previsible process γ_t, such that

$$\alpha(t,T) = \sigma(t,T)\left(\gamma_t - \Sigma(t,T)\right), \qquad \text{for all } t \leqslant T;$$

- the process $A_t = \Sigma(t,T)$ is non-zero for almost all (t,ω), $t < T$, for every maturity T;

- the expectation $\mathbb{E}\exp\frac{1}{2}\int_0^T \gamma_t^2\, dt$ is finite;

- and the expectation $\mathbb{E}\exp\frac{1}{2}\int_0^T (\gamma_t - \Sigma(t,T))^2\, dt$ is finite.

The importance of the first condition in this box is the constraint it places on the drift $\alpha(t,T)$. As the process γ_t is only a function of time and not of maturity, the drift is forced to take the value $-\sigma(t,T)\Sigma(t,T)$ modified only by the 'one-dimensional' displacement $\gamma_t\sigma(t,T)$. Given that $\sigma(t,T)$ and $\Sigma(t,T)$ are determined by the forward rate volatilities, the only degree of freedom for the drift comes from the one-parameter γ_t process. Unlike simple asset models, not all drift functions $\alpha(t,T)$ are allowable.

5.4 Short-rate models

Short-rate models are popular in the market. In particular, they are often used to price derivatives which depend only on one underlying bond. They have evolved from various historical starting points – some emerging from discrete frameworks, others from equilibrium models – and are often presented in a simple hierarchy with no apparent connection to any overarching model.

All however are HJM models, which is why we used this framework in the first place. And there is a mathematical transformation that makes these two alternative descriptions equivalent. Demonstrating that is the purpose of this section.

A short-rate model posits a risk-neutral measure $\mathbb{Q}$ and a short-rate process

r_t. The model is that instantaneous borrowing can take place at rate r_t for an infinitesimal period. Rolling up the periods gives rise to a cash bond process $B_t = \exp\left(\int_0^t r_s\,ds\right)$, as in the HJM model. As with the equations at the end of section 5.3, the bond prices are given by

$$P(t,T) = \mathbb{E}_{\mathbb{Q}}\left(\exp\left(-\int_t^T r_s\,ds\right)\bigm| \mathcal{F}_t\right),$$

and the value at time t of a claim X maturing at date T is

$$V_t = \mathbb{E}_{\mathbb{Q}}\left(\exp\left(-\int_t^T r_s\,ds\right)X \bigm| \mathcal{F}_t\right).$$

The paradigm of short-rate modelling is to work within a parameterised family of processes, which typically are Markovian. The parameters are chosen to best fit the market, and then the above expression for V_t is calculated to price the claim X.

HJM in terms of the short rate

It is not immediately clear that this is an HJM model. To prove this requires choosing the forward volatility surface $\sigma(t,T)$ so that the resulting short rate from the HJM model is exactly the same as the original process r_t. This is possible for any general short rate r_t, but it's easiest to show in the special case where r_t is a Markov process.

Suppose that that r_t is a Markov diffusion (though not necessarily time-homogeneous) with volatility $\rho(r_t, t)$ and drift $\nu(r_t, t)$. That is

$$dr_t = \rho(r_t, t)\,dW_t + \nu(r_t, t)\,dt,$$

where $\rho(x,t)$ and $\nu(x,t)$ are deterministic functions of space and time.

Then $\int_t^T f(t,u)\,du = -\log P(t,T) = g(r_t, t, T)$, where $g(x,t,T)$ is the deterministic function

$$g(x,t,T) = -\log\mathbb{E}_{\mathbb{Q}}\left(\exp\left(-\int_t^T r_s\,ds\right)\bigm| r_t = x\right).$$

There is a theorem:

Short-rate model in HJM terms

The required volatility structure is

$$\sigma(t,T) = \rho(r_t, t)\frac{\partial^2 g}{\partial x \partial T}(r_t, t, T),$$

and $\quad \Sigma(t,T) = -\rho(r_t, t)\frac{\partial g}{\partial x}(r_t, t, T).$

We can see why this is so, by thinking of the forward rate $f(t, T)$ as $\frac{\partial g}{\partial T}(r_t, t, T)$, and using Itô to deduce that

$$d_t f(t, T) = \frac{\partial^2 g}{\partial x \partial T}\left(\rho(r_t, t)\, dW_t + \nu(r_t, t)\, dt\right) + \frac{\partial^2 g}{\partial t \partial T}\, dt + \tfrac{1}{2}\frac{\partial^3 g}{\partial x^2 \partial T}\rho^2(r_t, t)\, dt.$$

The volatility term must match $\sigma(t, T)$, which gives us the result. In addition, the initial forward rate curve $f(0, T)$ is given by

$$f(0, T) = \frac{\partial g}{\partial T}(r_0, 0, T).$$

This volatility structure and initial curve then identifies an HJM model for this market with the same short rate under $\mathbb{Q}$.

Short rate in terms of HJM

Conversely, it is also true that HJM models are short-rate models. The equation for the bond price in terms of r_t holds (see near the end of section 5.3), with r_t in terms of the HJM volatilities $\sigma(t, T)$ and $\Sigma(t, T)$, given as

$$r_t = f(0, t) + \int_0^t \sigma(s, t)\, dW_s - \int_0^t \sigma(s, t)\Sigma(s, t)\, ds.$$

This formula is not necessarily simple.

Ho and Lee

Now for the accepted hierarchy of short-rate models: starting with Ho and Lee. In its short-rate form, Ho and Lee gives the SDE for r_t under $\mathbb{Q}$, the martingale measure, as

---

Ho and Lee model

The short rate is driven by the SDE:

$$dr_t = \sigma\, dW_t + \theta_t\, dt,$$

for some θ_t deterministic and bounded, and σ constant.

The question we should immediately ask is, which HJM model corresponds to this? Following the mechanics from earlier, we find via Itô

$$g(x, t, T) = x(T - t) - \tfrac{1}{6}\sigma^2(T - t)^3 + \int_t^T (T - s)\theta_s \, ds.$$

Thus $\sigma(t, T)$, the HJM volatility surface, is simply $\sigma \frac{\partial^2 g}{\partial x \partial T} = \sigma$. Thus the volatility surface is constant, depending on neither time nor maturity. We can fully specify the HJM model under $\mathbb{Q}$ as

Ho and Lee model in HJM terms

$$d_t f(t, T) = \sigma \, dW_t + \sigma^2(T - t) \, dt$$

$$\text{with} \quad f(0, T) = \frac{\partial g}{\partial T}(r_0, 0, T) = r_0 - \tfrac{1}{2}\sigma^2 T^2 + \int_0^T \theta_s \, ds.$$

Equivalently, we can provide the evolution of the bond prices $P(t, T)$ under $\mathbb{Q}$:

$$P(t, T) = \exp - \left(\sigma(T - t)W_t + \int_t^T f(0, u) \, du + \tfrac{1}{2}\sigma^2 T(T - t)t \right).$$

This model is the (general) single-factor model with constant volatility, and is actually the simple model of section 5.2. If used in the short-rate form, then σ sets the volatility of all forward rates and θ_t allows matching to any initial forward curve via the identity $f(0, T) = r_0 - \tfrac{1}{2}\sigma^2 T^2 + \int_0^T \theta_s \, ds$.

It is a simple model, and its simplicity tells against it – the forward rates and the short rate r_t can go negative occasionally, and go to infinity in the long term. And not just of course under $\mathbb{Q}$, but under any equivalent measure $\mathbb{P}$ as well. Many other models expend much effort just to avoid these pitfalls.

But it is not *that* simple a model – the HJM formulation allows a description of how the real forward curve can move over time. Given any previsible process γ_t, the forward rates can move as

$$d_t f(t, T) = \sigma \, dW_t + \left(\sigma^2(T - t) + \sigma\gamma_t \right) dt,$$
$$\text{with} \quad dr_t = \sigma \, dW_t + (\theta_t + \sigma\gamma_t) \, dt.$$

So short rates can have a wide range of possible drifts under $\mathbb{P}$ the real world measure, not just the simple deterministic drift θ_t. The restriction with Ho and Lee lies not there but in the implication that $\sigma(t, T) = \sigma$.

Two extra things need mentioning. First, the bond price and the cash bond price are both log-normally distributed and thus the Black–Scholes formula can still hold (as hinted at in section 4.1, and shown in section 6.2).

And second, there is a straightforward generalisation to a deterministic short-rate volatility

$$dr_t = \sigma_t \, dW_t + \theta_t \, dt,$$

with a corresponding HJM formulation

$$d_t f(t, T) = \sigma_t \, dW_t + \sigma_t^2 (T - t) \, dt,$$

with the initial forward rate curve given by

$$f(0, T) = r_0 - \int_0^T \sigma_s^2 (T - s) \, ds + \int_0^T \theta_s \, ds.$$

The extra freedom here is to allow the volatility surface to depend on time, but not on maturity. For that we require something else. . .

Vasicek/Hull–White

Next in the accepted hierarchy is to allow the short rate's drift to depend on its current value.

Vasicek model

We model the short rate (under $\mathbb{Q}$) as:

$$dr_t = \sigma \, dW_t + (\theta - \alpha r_t) \, dt$$

for some constant α, θ and σ.

The SDE is composed of a Brownian part and a restoring drift which pushes it upwards when the process is below θ/α and downwards when it is above. The magnitude of the drift is also proportional to the distance away from this mean. Such a process is called an Ornstein–Uhlenbeck or O-U process.

We can use Itô's formula to check that the solution to this, starting r at r_0, is

$$r_t = \theta/\alpha + e^{-\alpha t}(r_0 - \theta/\alpha) + \sigma e^{-\alpha t} \int_0^t e^{\alpha s}\, dW_s.$$

As it happens, r_t can be rewritten in terms of a different $\mathbb{Q}$-Brownian motion $\bar{W}$ as

$$r_t = \theta/\alpha + e^{-\alpha t}(r_0 - \theta/\alpha) + \sigma e^{-\alpha t} \bar{W}\left(\frac{e^{2\alpha t} - 1}{2\alpha}\right),$$

so that r_t has a normal marginal distribution with mean $\theta/\alpha + \exp(-\alpha t)(r_0 - \theta/\alpha)$ and variance $\sigma^2(1 - \exp(-2\alpha t))/2\alpha$. As t gets large, this converges to an equilibrium normal distribution of mean θ/α and variance $\sigma^2/2\alpha$. This does not mean that the process r_t converges – it doesn't – only that its distribution converges.

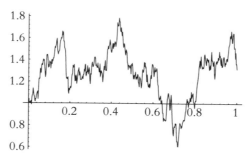

Figure 5.7 An O-U process, with $\sigma = 2\theta = 2\alpha = 1$

What HJM model are we in? Again we can use Itô to find $g(x, t, T)$, and thus $\sigma(t, T)$ and $f(0, T)$. In this case

Vasicek model in HJM terms

$$\sigma(t, T) = \sigma \exp\left(-\alpha(T - t)\right),$$

$$\text{with} \quad f(0, T) = \theta/\alpha + e^{-\alpha T}(r_0 - \theta/\alpha) - \frac{\sigma^2}{2\alpha^2}\left(1 - e^{-\alpha T}\right)^2.$$

Now we can see an advantage over Ho and Lee – where Ho and Lee failed to introduce a maturity dependence into the volatility surface, this model can. Thus this model is capable of calibration to a richer set of observed volatilities. Note how the volatility $\sigma(t, T)$ is derived from both the drift and volatility of the short rate under $\mathbb{Q}$. In order to describe an HJM model, we need two degrees of freedom for the volatility – one for time and one for maturity. The short rate description doesn't abandon the second degree of freedom; it encodes it in the relationship between its volatility $\rho(r_t, t)$ and its drift $\nu(r_t, t)$. The drift of r_t under $\mathbb{Q}$ is a vital part of the description.

But only under $\mathbb{Q}$. The Vasicek model, unlike Ho and Lee, may be mean reverting under $\mathbb{Q}$, but both models are in fact capable of mean reversion under $\mathbb{P}$. Some care has to be taken – the introduction of the extra parameter α does give Vasicek a richer set of allowable $\mathbb{P}$-drifts than Ho and Lee, but this richness involves maturity. Simple time-dependent considerations will not in general prejudice one over the other. Because it is possible to find a change of measure γ_t which gives mean reverting behaviour to Ho and Lee, Vasicek is not the inevitable choice if in the real world mean reversion is observed. In practice it will be the volatility of the entire curve, rather than the drift of the short rate that forces one over the other.

As before there is a natural generalisation to

$$dr_t = \sigma_t \, dW_t + (\theta_t - \alpha_t r_t) \, dt$$

where σ_t, θ_t, and α_t are deterministic functions of time. As r_t is still a Gaussian process with normal marginals, so $f(t, T)$ is Gaussian and the bond prices have log-normal marginals. In this case, the HJM volatility and initial forward curve are

$$\sigma(t, T) = \sigma_t \beta(t, T), \qquad \text{where} \quad \beta(t, T) = \exp\left(-\int_t^T \alpha_s \, ds\right), \qquad \text{and}$$

$$f(0, T) = r_0 \beta(0, T) + \int_0^T \theta_s \beta(s, T) \, ds - \int_0^T \sigma_s^2 \beta(s, T) \left(\int_s^T \beta(s, u) \, du\right) ds.$$

The normality of the forward rates $f(t, T)$ is both good news and bad news. In its favour, it means that the bond prices $P(t, T)$ are log-normally distributed, so that the log-normal option pricing results of section 6.2 all hold. On the other hand, both the instantaneous rate and the forward rates can go negative from time to time. Depending on the parameters, this can happen more or less rarely – the next model rectifies this defect.

Cox–Ingersoll–Ross

The model is a mean-reverting process, which pushes away from zero to keep it positive (see box).

Cox–Ingersoll–Ross model

The instantaneous rate's SDE, under $\mathbb{Q}$, is

$$dr_t = \sigma_t \sqrt{r_t}\, dW_t + (\theta_t - \alpha_t r_t)\, dt,$$

where σ_t, θ_t and α_t are deterministic functions of time.

The drift term is a restoring force which always points towards the current mean value of θ_t/α_t. The volatility term is set up to get smaller as r_t approaches zero, so allowing the drift θ_t to dominate and to stop r_t from going below zero.

As long as θ satisfies $\theta_t \geqslant \frac{1}{2}\sigma_t^2$, then the process actually stays strictly positive.

This process is called *autoregressive*.

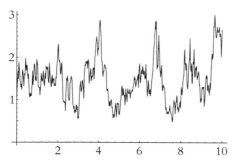

Figure 5.8 Autoregressive $\sigma = 1$, $\theta = 2$, $\alpha = 2$

It is difficult to find an explicit pathwise solution for r_t, but we can solve a useful partial differential equation (PDE). Firstly define $B(t,T)$ to be the solution of the Riccati differential equation

$$\frac{\partial B}{\partial t} = \tfrac{1}{2}\sigma_t^2 B^2(t,T) + \alpha_t B(t,T) - 1, \qquad B(T,T) = 0.$$

(In general, this equation has no analytic solution, but it has been well studied numerically.) Then the function $g(x, t, T)$, which is $(-\log P(t, T)|r_t = x)$ can be written in terms of this solution $B(t, T)$ as

$$g(x, t, T) = xB(t, T) + \int_t^T \theta_s B(s, T)\, ds.$$

Letting $D(t, T)$ be $\frac{\partial B}{\partial T}(t, T)$, then the volatility structure can be expressed as

Cox–Ingersoll–Ross in HJM terms

$$\sigma(t, T) = \sigma_t \sqrt{r_t}\, D(t, T),$$

$$\text{and} \qquad \Sigma(t, T) = -\sigma_t \sqrt{r_t}\, B(t, T),$$

$$\text{with} \qquad f(0, T) = r_0 D(0, T) + \int_0^T \theta_s D(s, T)\, ds.$$

As usual, the bond price $P(t, T)$ has the form $P(t, T) = \exp - g(r_t, t, T)$.

Black–Karasinski

Another way round the problem of keeping the short rate positive is to take exponentials. This model is an extension of the Black–Derman–Toy model and starts by taking X_t to be the general O–U process of the Vasicek model. Explicitly:

Black–Karasinski model
The process X_t is

$$dX_t = \sigma_t\, dW_t + (\theta_t - \alpha_t X_t)\, dt,$$

where σ_t, θ_t and α_t are deterministic functions of time. The instantaneous rate r_t is then assumed to be

$$r_t = \exp(X_t).$$

So the logarithm of the rate drifts towards the current mean of θ_t/α_t.

The rate itself also drifts towards a mean, and additionally is always positive. We also know that X_t is normally distributed, so that r_t is log-normal. However, $\int_t^T r_s \, ds$ is awkward to examine analytically. This model is still HJM consistent; that is, there is some volatility surface $\sigma(t,T)$ which generates a single-factor HJM model which has the same instantaneous rate as given above.

5.5 Multi-factor HJM

The drawback of the single-factor model is that all the increments in the bond prices are perfectly correlated. For many applications, that assumption is too coarse, especially if we are trying to price something which depends on the difference of two points on the yield curve.

A multi-factor model involves driving the various processes by a collection of independent Brownian motions. More details of such models are in section 6.3. In an n-factor model, we will have n Brownian motions to work with: $W_1(t), \ldots, W_n(t)$. Correspondingly each T-bond forward rate process has a volatility $\sigma_i(t,T)$ for each Brownian factor $W_i(t)$. This allows different bonds to depend on external 'shocks' in different ways, and to have strong correlations with some bonds and weaker correlations with others. The general form of the multi-factor HJM model is

$$f(t,T) = f(0,T) + \sum_{i=1}^{n} \int_0^t \sigma_i(s,T) \, dW_i(s) + \int_0^t \alpha(s,T) \, ds, \quad 0 \leqslant t \leqslant T,$$

which is to say that the forward process starts with initial value $f(0,T)$ and is driven by various Brownian terms and a drift. From this, the total instantaneous square volatility of $f(t,T)$, and and the covariance of the increments of the two forward rates $f(t,T)$ and $f(t,S)$ are respectively

$$\sum_{i=1}^{n} \sigma_i^2(t,T), \quad \text{and} \quad \sum_{i=1}^{n} \sigma_i(t,T)\sigma_i(t,S).$$

In the single-factor model, n is 1, and the correlation of the changes in the forward rates of the T-bond and S-bond is exactly one.

5.5 Multi-factor HJM

The instantaneous rate $r_t = f(t, t)$ can be written, similarly to before as

$$r_t = f(0, t) + \sum_{i=1}^{n} \int_0^t \sigma_i(s, t) \, dW_i(s) + \int_0^t \alpha(s, t) \, ds.$$

The volatility and drift conditions are generalised to:

Multi-factor HJM: conditions on the volatilities and drift

We assume that

- for each T, the processes $\sigma_i(t, T)$ and $\alpha(t, T)$ are $\mathcal{F}$-previsible and their integrals $\int_0^T \sigma_i^2(t, T) \, dt$ and $\int_0^T |\alpha(t, T)| \, dt$ are finite;

- the initial forward curve, $f(0, T)$, is deterministic and satisfies the condition that $\int_0^T |f(0, u)| \, du < \infty$;

- the drift α has finite integral $\int_0^T \int_0^u |\alpha(t, u)| \, dt \, du$;

- each volatility σ_i has finite expectation $\mathbb{E} \int_0^T \left| \int_0^u \sigma_i(t, u) \, dW_i(t) \right| du$.

To make the discounted bond prices into martingales, we need a version of the Cameron–Martin–Girsanov theorem for higher dimensions (section 6.3). The conditions we need for this to work are shown in the two boxes, where $\Sigma_i(t, T)$ is the integral $-\int_t^T \sigma_i(t, u) \, du$.

Multi-factor HJM: market completeness conditions (1)

It is required that

- there exist previsible processes $\gamma_i(t)$, for $1 \leqslant i \leqslant n$, such that

$$\alpha(t, T) = \sum_{i=1}^{n} \sigma_i(t, T) \big(\gamma_i(t) - \Sigma_i(t, T) \big), \qquad \text{for all } t \leqslant T;$$

- the expectation $\mathbb{E} \exp \frac{1}{2} \sum_i^n \int_0^T \gamma_i^2(t) \, dt$ is finite;

This is just as in the single-factor case, but with one difference. The drift is now allowed n 'dimensions of freedom' away from its risk–neutral value. That is, as a function of T, $\alpha(t, \cdot)$ is allowed to deviate by any linear combination of the functions $\sigma_i(t, \cdot)$. This is still much less than the set of all possible functions, but it is larger than in the single-factor case. The second condition is the technical requirement of the C–M–G theorem for $\gamma_i(t)$ to be a drift under an equivalent change of measure.

Multi-factor HJM: market completeness conditions (2)

We also need that

- the matrix $A_t = \left(\Sigma_i(t, T_j)\right)^n_{i,j=1}$ is non-singular for almost all (t, ω), $t < T_1$, for every set of maturities $T_1 < T_2 < \ldots < T_n$;

- and the expectation $\mathbb{E} \exp \frac{1}{2} \sum_i^n \int_0^T \left(\gamma_i(t) - \Sigma_i(t, T)\right)^2 dt$ is finite.

The modification from the single-factor case here is that the volatility process A_t which used to be required to be non-zero has been replaced by a volatility matrix process which has to be non-singular. The second condition ensures that the resulting driftless discounted bond price is in fact a martingale (a multi-dimensional equivalent of the collector's guide to exponential martingales).

As before we find that the bond prices themselves have stochastic increments

$$
d_t P(t, T) = P(t, T) \left(\sum_{i=1}^n \Sigma_i(t, T) \, dW_i(t) \ldots \right.
$$

$$
\left. + \left(r_t - \int_t^T \left(\alpha(t, u) + \sum_{i=1}^n \sigma_i(t, u)\Sigma_i(t, u)\right) du \right) dt \right),
$$

where $\Sigma_i(t, T)$ is the integral $-\int_t^T \sigma_i(t, u) \, du$. The discounted bond prices $Z(t, T) = B_t^{-1} P(t, T)$ satisfy

$$
d_t Z(t, T) = Z(t, T) \left(\sum_{i=1}^n \Sigma_i(t, T) \, dW_i(t) \right.
$$

$$
\left. - \left(\int_t^T \left(\alpha(t, u) + \sum_i \sigma_i(t, u)\Sigma_i(t, u)\right) du \right) dt \right).
$$

The SDE for Z now becomes

$$d_t Z(t,T) = Z(t,T) \sum_{i=1}^{n} \Sigma_i(t,T) \big(dW_i(t) + \gamma_i(t)\,dt\big).$$

Using the multi-dimensional C-M-G, we can find a measure $\mathbb{Q}$ equivalent to $\mathbb{P}$, under which $\tilde{W}_1, \ldots, \tilde{W}_n$ are independent $\mathbb{Q}$-Brownian motions, where $\tilde{W}_i(t) = W_i(t) + \int_0^t \gamma_i(s)\,ds$. So Z's SDE is (in $\mathbb{Q}$-terms)

$$d_t Z(t,T) = Z(t,T) \sum_{i=1}^{n} \Sigma_i(t,T)\,d\tilde{W}_i(t),$$

and every $Z(t,T)$ is a $\mathbb{Q}$-martingale in t.

Under this martingale measure, the bond price P and the forward rate f have the stochastic differentials

Bond prices and forward rates under $\mathbb{Q}$

$$d_t P(t,T) = P(t,T) \left(\sum_{i=1}^{n} \Sigma_i(t,T)\,d\tilde{W}_i(t) + r_t\,dt \right),$$

$$d_t f(t,T) = \sum_{i=1}^{n} \sigma_i(t,T)\,d\tilde{W}_i(t) - \sum_{i=1}^{n} \sigma_i(t,T)\Sigma_i(t,T)\,dt.$$

Derivative pricing and hedging

The actual price of a derivative still has a familiar form:

Option price formula (HJM)

If X is the payoff of a derivative at time T, then its value at time t is

$$V_t = B_t \, \mathbb{E}_{\mathbb{Q}}\big(B_T^{-1} X \mid \mathcal{F}_t\big) = \mathbb{E}_{\mathbb{Q}}\Big(\exp\big(-\textstyle\int_t^T r_s\,ds\big) X \mid \mathcal{F}_t\Big).$$

We also need a multi-dimensional martingale representation theorem. Formally

Martingale representation theorem (n–factor)
Let $\tilde{W}$ be n–dimensional $\mathbb{Q}$–Brownian motion, and suppose that M_t is an n–dimensional $\mathbb{Q}$–martingale process, $M_t = (M_1(t), \ldots, M_n(t))$, which has volatility matrix $(\sigma_{ij}(t))$, in that $dM_j(t) = \sum_i \sigma_{ij}(t)\, d\tilde{W}_i(t)$, and the matrix satisfies the additional condition that (with probability one) it is always non-singular. Then if N_t is any one-dimensional $\mathbb{Q}$–martingale, there exists an n–dimensional $\mathcal{F}$–previsible process $\phi_t = (\phi_1(t), \ldots, \phi_n(t))$ such that $\int_0^T \left(\sum_j \sigma_{ij}(t)\phi_j(t)\right)^2 dt < \infty$, and the martingale N can be written as

$$N_t = N_0 + \sum_{j=1}^{n} \int_0^t \phi_j(s)\, dM_j(s).$$

Further ϕ is (essentially) unique.

As a general rule, if we have an n–factor model, we need a trading portfolio of n separate instruments, as well as the cash bond, in order to hedge claims. An advantage of the HJM framework is that we are free to choose whichever n instruments we like, and the answer will always be the same.

If we are going to hedge the claim X with discount bonds, we must still make sure that all their maturities are later than T. Suppose we choose to use bonds with maturities $T_1, T_2, \ldots, T_n$ all larger than T.

A self-financing strategy $(\phi_1(t), \ldots, \phi_n(t), \psi_t)$ will be the combination of both an n–vector of holdings in the bonds with maturities $T_1, \ldots, T_n$ respectively, and a holding ψ_t in the cash bond B_t. The value of the portfolio at time t is

$$V_t = \sum_{j=1}^{n} \phi_j(t) P(t, T_j) + \psi_t B_t,$$

and its discounted value $E_t = B_t^{-1} V_t$ is

$$E_t = \sum_{j=1}^{n} \phi_j(t) Z(t, T_j) + \psi_t.$$

The self-financing equality for the strategy (as in section 6.4) is that

$$dE_t = \sum_{j=1}^{n} \phi_j(t)\, d_t Z(t, T_j).$$

We can now apply the representation theorem in the usual way to the martingale produced from the discounted claim, that is $E_t = \mathbb{E}_{\mathbb{Q}}(B_T^{-1}X|\mathcal{F}_t)$. The part of the martingales $M_j(t)$ will be taken by the discounted bonds $Z(t, T_j)$. Their volatility matrix is given by $A_t = (\Sigma_i(t, T_j))_{i,j}$, which is non-singular by the completeness conditions box. If we set $E_t = \mathbb{E}_{\mathbb{Q}}(B_T^{-1}X|\mathcal{F}_t)$, then by the representation theorem, there is an n-vector of previsible processes ϕ_t such that

$$E_t = \mathbb{E}_{\mathbb{Q}}(B_T^{-1}X) + \sum_{j=1}^{n} \int_0^t \phi_j(s)\, dZ(s, T_j).$$

This immediately gives a self-financing strategy ϕ. We hold $\phi_j(t)$ units of the T_j-bond at time t, and $\psi_t = E_t - \sum_j \phi_j(t)Z(t, T_j)$ units of the cash bond.

In the usual way, the portfolio costs an initial $\mathbb{E}_{\mathbb{Q}}(B_T^{-1}X)$ and evolves to be worth exactly X by time T.

5.6 Interest rate products

In recent years, there has been a great increase in the number of interest rate products available. Especially in the over-the-counter markets, contracts which not long ago would have been considered as exotics are now commonplace. We cannot hope to describe the hundreds and possibly thousands of traded claims, but we can sketch out the basic types within each area.

Forward contract

This is about the simplest product possible. We agree, at the current time t, to make a payment of an amount k at a future time T_1, and in return to receive a dollar at the later time T_2. What should the amount k be?

According to the pricing formula (under whatever model we are in), the value of the claim now is

$$V_t = B_t\mathbb{E}_{\mathbb{Q}}(B_{T_2}^{-1}|\mathcal{F}_t) - B_t\mathbb{E}_{\mathbb{Q}}(kB_{T_1}^{-1}|\mathcal{F}_t),$$

under the martingale measure $\mathbb{Q}$, where B_t is the cash bond

$$B_t = \exp \int_0^t r_s\, ds.$$

Recalling that $B_t \mathbb{E}_{\mathbb{Q}}(B_T^{-1}|\mathcal{F}_t)$ is just $P(t,T)$, we see that

$$V_t = P(t, T_2) - kP(t, T_1).$$

For this contract to have null current net worth, we merely choose k at time t to be

$$k = \frac{P(t, T_2)}{P(t, T_1)}.$$

This price makes sense, as saying that the forward yield from T_1 to T_2 is

$$-\frac{\log P(t, T_2) - \log P(t, T_1)}{T_2 - T_1}.$$

For T_1 and T_2 very close together, this approximates to the instantaneous forward rate of borrowing

$$-\frac{\partial}{\partial T} \log P(t, T) = f(t, T).$$

The price also gives us a clue to the hedging strategy. Suppose we were, at time t, to buy k units of the T_1-bond and sell one unit of the T_2-bond. The initial cost of that deal is zero, and the portfolio pays us k at time T_1 (matching the payment we have to make at that time) and exactly absorbs the dollar we receive at time T_2.

In this particular example, the answer is independent of our particular term structure model, as the hedging strategy is static. There are other important cases where this also happens.

Multiple payment contracts

Most interest rate products don't just make a single payment X at time T. Instead the contract specifies a sequence of payments X_i made at a sequence of times T_i ($i = 1, \ldots, n$). Each payment X_i may depend on price movements up to its payment time T_i, and even on any previous payment. As long as we bear that in mind, this causes no serious problem, and indeed there are two different ways to keep things clear.

- **Divide and rule.** We can treat each payment X_i separately. On its own, it is just a claim at time T_i, so its worth at time t is

$$V_i(t) = B_t \mathbb{E}_{\mathbb{Q}}\big(B_{T_i}^{-1} X_i \mid \mathcal{F}_t\big) = P(t, T_i)\mathbb{E}_{\mathbb{P}_{T_i}}(X|\mathcal{F}_t),$$

where $\mathbb{P}_{T_i}$ is the T_i-forward measure (see section 6.4). This approach will always work, but the forward measure, if used, will have to be changed for each i.

- **Savings account.** We could instead roll up the payments into savings as we get them, and keep them till the last payment date T. That is, as each payment is made, we use it to buy a T-bond (or invest it in the bank account process B_t till time T). Then the payoff is a single payment at time T of

$$X = \sum_{i=1}^{n} \frac{X_i}{P(T_i, T)}$$

with worth at time t

$$V_t = B_t \mathbb{E}_{\mathbb{Q}}\left(B_T^{-1} X \mid \mathcal{F}_t\right) = P(t, T)\mathbb{E}_{\mathbb{P}_T}(X|\mathcal{F}_t).$$

Bonds with coupons

In practice, pure discount bonds with no coupon are not popular products. Especially at the long end. Instead, a bond may not only pay its principal back at maturity, but also make smaller regular coupon payments of a fixed amount c up until then.

Suppose a bond makes n regular payments at (uncompounded) rate k at times $T_i = T_0 + i\delta$ $(i = 1, \ldots, n)$ and also pays off a dollar at time T_n. The amount of the actual coupon payment is $k\delta$, where δ is the payment period. This income stream is equivalent to owning one T_n-bond and $k\delta$ units of each T_i-bond. The price of the coupon bond at time T_0 is

$$P(T_0, T_n) + k\delta \sum_{i=1}^{n} P(T_0, T_i).$$

If we desire the bond to start with its face value, then the coupon rate should be

$$k = \frac{1 - P(T_0, T_n)}{\delta \sum_{i=1}^{n} P(T_0, T_i)}.$$

Floating rate bonds

A bond might also pay off a coupon which was not fixed, but depended on current interest rates. One interesting case is where the interest paid over an interval from time S to time T is the same as the yield of the T-bond bought at time S.

Suppose a bond pays its dollar principal at time T_n, and also payments at times $T_i = T_0 + i\delta$ $(i = 1, \ldots, n)$ of varying amounts. The amount of payment made at time T_i is determined by the LIBOR rate set at time T_{i-1}

$$L(T_{i-1}) = \frac{1}{\delta}\left(\frac{1}{P(T_{i-1}, T_i)} - 1\right).$$

The actual payment made at time T_i is $\delta L(T_{i-1}) = P(T_{i-1}, T_i)^{-1} - 1$, which is the amount of interest we would receive by buying a dollar's worth of the T_i bond at time T_{i-1}.

The value to us now, at time T_0, of the T_i payment is

$$B_{T_0}\,\mathbb{E}_\mathbb{Q}\big(B_{T_i}^{-1}(P^{-1}(T_{i-1}, T_i) - 1) \mid \mathcal{F}_{T_0}\big).$$

Because the conditional expectation $\mathbb{E}_\mathbb{Q}(B_{T_i}^{-1}|\mathcal{F}_{T_{i-1}})$ is $B_{T_{i-1}}^{-1}P(T_{i-1}, T_i)$, and the bond price $P(T_{i-1}, T_i)$ is known with respect to the $\mathcal{F}_{T_{i-1}}$-information, we can divide it through both sides to get

$$\mathbb{E}_\mathbb{Q}\big(B_{T_i}^{-1}P^{-1}(T_{i-1}, T_i) \mid \mathcal{F}_{T_{i-1}}\big) = B_{T_{i-1}}^{-1}.$$

Using the tower law, we can rewrite the value of the T_i payment as

$$B_{T_0}\,\mathbb{E}_\mathbb{Q}\big(B_{T_{i-1}}^{-1} - B_{T_i}^{-1} \mid \mathcal{F}_{T_0}\big),$$

which is just $P(T_0, T_{i-1}) - P(T_0, T_i)$. This price also suggests the hedge of selling a T_i-bond and buying a T_{i-1}-bond. When the T_{i-1}-bond matures, we buy $P^{-1}(T_{i-1}, T_i)$ units of the T_i-bond, and we are left with exactly the right payoff at time T_i.

The total value of the variable coupon bond is the sum of its components. That is,

$$V_0 = P(T_0, T_n) + \sum_{i=1}^{n}\big(P(T_0, T_{i-1}) - P(T_0, T_i)\big) = 1.$$

Surprisingly, the bond has a fixed price equal to the face value of its principal. Why this is so, is because the bond is equivalent to this simple sequence of trades:

- take a dollar and buy T_1-bonds with it

- take the interest from the bonds at T_1 as a coupon, and buy some T_2-bonds with the dollar principal

- repeat until we are left with the dollar at time T_n.

This has exactly the same cash flows as the variable coupon bond, so the initial prices must match.

Swaps

This very popular contract simply exchanges a stream of varying payments for a stream of fixed amount payments (or *vice versa*). That is, we swap a floating interest rate for a fixed one.

Typically, we might offer a contract where we receive a regular sequence of fixed amounts and at each payment date we pay an amount depending on prevailing interest rates. In practice, only the net difference is exchanged, as shown in figure 5.9:

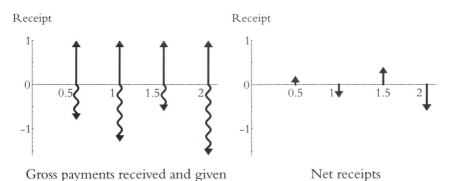

Gross payments received and given   Net receipts

Figure 5.9

A standard definition of the variable payment is that of the interest paid by a bond over the previous time period. If the payment dates are $T_i = T_0 + i\delta$ ($i = 1, \ldots, n$), then the ith payment will be determined by the δ-period

LIBOR rate set at time T_{i-1}. The payment made is

$$\delta L(T_{i-1}) = \frac{1}{P(T_{i-1}, T_i)} - 1.$$

Suppose the swap pays at a fixed rate k at each time period. Then the swap looks like a portfolio which is long a fixed coupon bond and short a variable coupon bond. We know that the former is worth

$$P(T_0, T_n) + k\delta \sum_{i=1}^{n} P(T_0, T_i),$$

and the latter costs a dollar. The fixed rate needed to give the swap initial null value is

$$k = \frac{1 - P(T_0, T_n)}{\delta \sum_{i=1}^{n} P(T_0, T_i)}.$$

Forward swaps

In a forward swap agreement, we have chosen to receive fixed payments at rate k, starting at time T_0 with payments at times $T_i = T_0 + i\delta$ $(i = 1, \ldots, n)$. The value of this swap at time T_0 will be

$$X = P(T_0, T_n) + k\delta \sum_{i=1}^{n} P(T_0, T_i) - 1.$$

The present value of X at time t before T_0 is given by the formula:

$$V_t = B_t \mathbb{E}_{\mathbb{Q}}\left(B_{T_0}^{-1} X \mid \mathcal{F}_t\right) = P(t, T_n) + k\delta \sum_{i=1}^{n} P(t, T_i) - P(t, T_0).$$

The fixed rate needed to give the forward swap initial null value at time t is

$$k = \frac{P(t, T_0) - P(t, T_n)}{\delta \sum_{i=1}^{n} P(t, T_i)}.$$

This rate k is the *forward swap rate*. An alternative formulation of this expression is

$$k = \frac{1 - F_t(T_0, T_n)}{\delta \sum_{i=1}^{n} F_t(T_0, T_i)},$$

where $F_t(T_0, T_i)$ is the forward price at time t for purchasing a T_i-bond at time T_0. That is $F_t(T_0, T_i) = P(t, T_i)/P(t, T_0)$. In this form the expression resembles the instantaneous swap rate.

Bond options

Like a stock option, a bond option gives the right to buy a bond at a future date for a given price. An option on a T-bond, struck at k with exercise time t, has current worth

$$\mathbb{E}_{\mathbb{Q}}\left(B_t^{-1}\left(P(t,T)-k\right)^+\right),$$

where $\mathbb{Q}$ is the martingale measure.

Under the Ho and Lee model, where the forward rates evolve as

$$d_t f(t,T) = \sigma\,dW_t + \sigma^2(T-t)\,dt,$$

the forward rates and the instantaneous short rate are normally distributed. This makes the T-bond and the discount bond log-normally distributed, so that we can price the option with the log-normal results of section 6.2. The option price is

$$V_0 = P(0,t)\left(F\Phi\left(\frac{\log\frac{F}{k}+\frac{1}{2}\bar{\sigma}^2 t}{\bar{\sigma}\sqrt{t}}\right) - k\Phi\left(\frac{\log\frac{F}{k}-\frac{1}{2}\bar{\sigma}^2 t}{\bar{\sigma}\sqrt{t}}\right)\right),$$

where F is the current forward price for $P(t,T)$, that is $F = P(0,T)/P(0,t)$, and the *term volatility* $\bar{\sigma}$ is $\sigma(T-t)$ (that is, $\bar{\sigma}^2 t$ is the log-variance of $P(t,T)$). Under the Vasicek model, which is the most general single-factor model with log-normal bond prices, this formula also holds with the same forward price, but a different $\bar{\sigma}$ depending on the deterministic processes σ_t and ϕ_t in the model.

Compare this with the price of an option on a stock S, with volatility σ, struck at price k with exercise time t. It is worth

$$V_0 = e^{-rt}\left(F\Phi\left(\frac{\log\frac{F}{k}+\frac{1}{2}\sigma^2 t}{\sigma\sqrt{t}}\right) - k\Phi\left(\frac{\log\frac{F}{k}-\frac{1}{2}\sigma^2 t}{\sigma\sqrt{t}}\right)\right),$$

where r is the constant interest rate and F is the current forward price of the stock, that is $F = e^{rt}S_0$, and σ is the (term) volatility of S_t.

We see that the bond option price formula merely changes the discount factor representing the value now of a dollar at time t. Under constant interest rates this was e^{rt}, and under variable interest rates it is just the price of a t-bond $P(0,t)$. Otherwise, as long as the other variables are expressed in terms of forward prices and term volatilities, the formula is the same.

Options on coupon bonds

Imagine a bond which pays coupons at rate k at the times $T_i = T_0 + i\delta$ $(i = 1, \ldots, n)$ before redeeming a dollar at time T_n. We can buy or sell the bond before time T_n, transferring the ownership of future (but not past) coupons along with it. As we've seen before the value of this bond at time t is

$$C_t = P(t, T_n) + k\delta \sum_{i=I(t)}^{n} P(t, T_i),$$

where $I(t) = \min\{i : t < T_i\}$ is the sequence number of the next coupon payment after time t.

Suppose we have an option to buy the bond at time t for price K. In general it is not easy to value this option analytically. However, in the special case where we have a single-factor model with a Markovian short rate, we can price the option more easily using a trick of Jamshidian.

Each bond price $P(t, T)$ can be seen as a deterministic function $P(t, T; r_t)$ of time, maturity and the instantaneous rate. Additionally, this function will be decreasing in r_t — as rates rise, prices fall. A portfolio which is long a number of bonds will have the same behaviour. So C_t itself will be a function $C(t; r_t)$ which is decreasing in r_t.

Thus there is some critical value r^* of r such that $C(t; r^*)$ is exactly K. Setting K_i to be $P(t, T_i; r^*)$, then r^* is also critical for an option on the T_i-bond struck at K_i. This means that C_t is larger than K if and only if any (and every) $P(t, T_i)$ is larger than K_i. And so

$$(C_t - K)^+ = \left(P(t, T_n) - K_n\right)^+ + k\delta \sum_{i=I(t)}^{n} \left(P(t, T_i) - K_i\right)^+.$$

In other words, an option on this portfolio is a portfolio of options, and we can price each one using the zero-coupon bond option formula.

Caps and floors

Suppose we are borrowing at a floating rate and want to insure against interest payments going too high. If we make payments at times $T_i = T_0 + i\delta$ $(i = 1, \ldots, n)$, then we pay at time T_i the δ-period LIBOR rate set at time T_{i-1}

$$L(T_{i-1}) = \frac{1}{\delta} \left(\frac{1}{P(T_{i-1}, T_i)} - 1 \right).$$

How much would it cost to ensure that this rate is never greater than some fixed rate k? The cap contract pays us the difference between the LIBOR and the cap rate

$$\delta\left(L(T_{i-1}) - k\right)^{+}$$

at each time T_i. An individual payment at a particular time T_i is called a *caplet*, and if we can price caplets, we can price the cap.

Now we can rewrite the caplet claim as

$$X = (1 + k\delta)P_i^{-1}\left(K - P_i\right)^{+},$$

where P_i is $P(T_{i-1}, T_i)$ and K is $(1 + k\delta)^{-1}$. The value of the caplet at time t is $B_t\mathbb{E}_{\mathbb{Q}}(B_{T_i}^{-1}X|\mathcal{F}_t)$, which equals

$$(1 + k\delta)\, B_t\mathbb{E}_{\mathbb{Q}}\left(B_{T_{i-1}}^{-1}\left(K - P_i\right)^{+}\,\Big|\,\mathcal{F}_t\right).$$

This is just equal to the value of $(1 + k\delta)$ put options on the T_i-bond, struck at K, exercised at T_{i-1}. The option price formula (and put–call parity) will then price the caplet.

A floor works similarly, but inversely, in that we receive a premium for agreeing to never pay less than rate k at each time T_i. That is, we pay an extra amount

$$\delta\left(k - L(T_{i-1})\right)^{+}$$

at time T_i. There is a floor–cap parity which says that the worth of a 'floorlet' less the cost of a caplet equals $(1 + k\delta)P(t, T_i) - P(t, T_{i-1})$. Buying a floor and selling a cap at the same strike k is exactly equivalent to receiving fixed at rate k on a swap.

Swaptions

A *swaption* is an option to enter into a swap on a future date at a given rate. Suppose we have an option to receive fixed on a swap starting at date T_0. The swap payment dates are $T_i = T_0 + i\delta$ $(i = 1, \ldots, n)$, and the fixed swap rate is k. Then the worth of the option at time T_0 is

$$\left(P(T_0, T_n) + k\delta \sum_{i=1}^{n} P(T_0, T_i) - 1\right)^{+}.$$

This is exactly the same as a call option, struck at 1, on a T_n-bond which pays a coupon at rate k at each time T_i. That is not entirely a coincidence as a swap is just a coupon bond less a floating bond (which always has par value). If you receive fixed on a swap, you have a long position in the bond market; a swap option looks like a bond option.

5.7 Multi-factor models

If we want to price a product depending on a range of bonds, it makes more sense to use a multi-factor model. A simple case is given in Heath–Jarrow–Morton's original paper. It is an extension of Ho and Lee's model to two factors.

A two-factor model

Suppose the forward rates evolve as

$$d_t f(t,T) = \sigma_1\, dW_1(t) + \sigma_2 e^{-\lambda(T-t)}\, dW_2(t) + \alpha(t,T)\, dt,$$

where σ_1, σ_2 and λ are constants, and α is a deterministic function of t and T. Here the W_1 Brownian motion provides 'shocks' which are felt equally by points of all maturities on the yield curve, whereas W_2 gives short-term shocks which have little effect on the long-term end of the curve. This model is HJM consistent, so we can read off information about it from that structure. The HJM completeness conditions reduce, in this case, to there being two $\mathcal{F}$-previsible processes $\gamma_1(t)$ and $\gamma_2(t)$ such that the drift α is

$$\alpha(t,T) = \sigma_1\gamma_1(t) + \sigma_2 e^{-\lambda(T-t)}\gamma_2(t) + \sigma_1^2(T-t) + \frac{\sigma_2^2}{\lambda}\left(1 - e^{-\lambda(T-t)}\right)e^{-\lambda(T-t)}.$$

So the range of available drifts has two degrees of functional freedom away from the martingale measure drift. Under the martingale measure (that is $\gamma_1 = \gamma_2 = 0$), the forward rate is

$$f(t,T) = \sigma_1 W_1(t) + \sigma_2 e^{-\lambda T}\int_0^t e^{\lambda s}\, dW_2(s) + f(0,T) + \int_0^t \alpha(s,T)\, ds.$$

Like Ho and Lee, this model has normally distributed forward rates – which does allow them to go negative. Nevertheless the model does have the advantages of technical tractability and an explicit option formula. We can deduce from the forward rate formula that $-\log P(t,T) = \int_t^T f(t,u)\,du$ is

$$
\sigma_1(T-t)W_1(t) + \frac{\sigma_2}{\lambda}\left(e^{-\lambda t} - e^{-\lambda T}\right)\int_0^t e^{\lambda s}\,dW_2(s)
$$
$$
+ \int_t^T f(0,u)\,du + \int_0^t \int_t^T \alpha(s,u)\,du\,ds,
$$

and that the instantaneous interest rate is

$$
r_t = \sigma_1 W_1(t) + \sigma_2 e^{-\lambda t}\int_0^t e^{\lambda s}\,dW_2(s) + f(0,t) + \int_0^t \alpha(s,t)\,ds.
$$

This means that the instantaneous rate is made up of a Brownian motion and an independent mean-reverting (Ornstein–Uhlenbeck) process plus drift. However in a multi-factor setting, the short rate loses its dominant role as the carrier of all information about the bond prices.

Setting $\bar{\sigma}^2(t,T)$ to be the variance (term variance) of $\log P(t,T)$, we have

$$
\bar{\sigma}^2(t,T) = \sigma_1^2(T-t)^2 t + \left(\frac{\sigma_2}{\lambda}\left(1 - e^{-\lambda(T-t)}\right)\right)^2 \frac{1}{2\lambda}\left(1 - e^{-2\lambda t}\right).
$$

The discounted bond, $B_t = \exp(\int_0^t r_s\,ds)$, is also log-normally distributed, because we can deduce that the integral $\int_0^t r_s\,ds$ is normal from the expression for r_t above. We can use the results of section 6.2, given the joint log-normality of the asset and discount bond prices. The value of an option on the T-bond, struck at k, exercised at time t is

$$
V_0 = P(0,t)\left(F\Phi\left(\frac{\log\frac{F}{k} + \frac{1}{2}\bar{\sigma}^2(t,T)}{\bar{\sigma}(t,T)}\right) - k\Phi\left(\frac{\log\frac{F}{k} - \frac{1}{2}\bar{\sigma}^2(t,T)}{\bar{\sigma}(t,T)}\right)\right),
$$

where F is $P(0,T)/P(0,t)$, the forward price of the T-bond. This Black–Scholes type of formula allows us to price caps and floors as well as options on the discount T-bonds. However, in the multi-factor setting, the trick we used before to price options on coupon-bearing bonds does not work, making it more involved to price them and the associated swaptions.

The general multi-factor normal model

We can actually generalise the two-factor model above to a general multi-factor one which also has normal forward rates and an explicit Black–Scholes type option pricing formula.

We take the instance of the completely general n-factor model, where each volatility surface $\sigma_i(t,T)$ can be written as a product

$$\sigma_i(t,T) = x_i(t)y_i(T),$$

where x_i and y_i are deterministic functions. The forward rates are then driven by

$$d_t f(t,T) = \sum_{i=1}^{n} y_i(T)x_i(t)\,dW_i(t) + \alpha(t,T)\,dt.$$

Here the function x_i determines the size at time t of 'type i shocks', and the function y_i controls how the shock is felt at different maturities. In the single-factor case when $n = 1$, this framework incorporates both the Ho and Lee model ($x(t) = \sigma$, $y(T) = 1$) and the Vasicek model ($x(t) = \sigma_t \exp\left(\int_0^t \alpha_s\,ds\right)$, $y(T) = \exp\left(-\int_0^T \alpha_s\,ds\right)$).

For the market to be complete, we need two conditions on the functions α and y_i to hold. Firstly, there should be n $\mathcal{F}$-previsible processes $\gamma_1, \ldots, \gamma_n$, such that

$$\alpha(t,T) = \sum_{i=1}^{n} x_i(t)y_i(t)\big(\gamma_i(t) + x_i(t)Y_i(t,T)\big),$$

where $Y_i(t,T) = \int_t^T y_i(u)\,du$. In other words, the drifts consistent with hedging span an n-dimensional function space around the martingale drift. Secondly the matrix $A_t = \big(a_{ij}(t)\big)$, where $a_{ij}(t) = Y_j(t,T_i)$ should be non-singular for all $t < T_1$, for every set of n maturities $T_1 < \ldots < T_n$. This condition is really just asserting that all the functions y_i are different. It is satisfied, for instance, if each volatility σ_i has the form

$$\sigma_i(t,T) = \sigma_i(t)\exp\big(-\lambda_i(T-t)\big),$$

where the $\sigma_i(t)$ are deterministic functions of time and the λ_i are distinct constants.

For the general volatility surface $\sigma_i(t,T) = x_i(t)y_i(T)$, the short rate and the forward rates are normally distributed. Consequently the bond prices are log-normally distributed and a Black–Scholes type formula holds (see section 6.2). Let F be the forward price of the T-bond at time t, $F = P(0,T)/P(0,t)$, and let σ be the term volatility of the T-bond up to time t, that is $\sigma^2 t$ is the variance of $\log P(t,T)$, or

$$\sigma^2 = \frac{1}{t} \sum_{i=1}^{n} Y_i^2(t,T) \int_0^t x_i^2(s) \, ds.$$

Then the value at time zero of a call on the T-bond, struck at k, exercisable at time t is

$$V_0 = P(0,t) \left(F\Phi \left(\frac{\log \frac{F}{k} + \frac{1}{2}\sigma^2 t}{\sigma\sqrt{t}} \right) - k\Phi \left(\frac{\log \frac{F}{k} - \frac{1}{2}\sigma^2 t}{\sigma\sqrt{t}} \right) \right).$$

Brace–Gatarek–Musiela

The Brace–Gatarek–Musiela (BGM) model is a particular case of HJM which focuses on the δ-period LIBOR rates. We shall simplify their notation slightly and write

$$L(t,T) = \frac{1}{\delta} \left(\frac{P(t,T)}{P(t,T+\delta)} - 1 \right).$$

So $L(t,T)$ is the δ-period (forward) LIBOR rate for borrowing at a time T.

The general HJM model (of n factors) defined by the forward volatilities $\sigma_i(t,T)$ is restricted in the BGM setup to those σ such that

$$\int_T^{T+\delta} \sigma_i(t,u) \, du = \frac{\delta L(t,T)}{1 + \delta L(t,T)} \gamma_i(t,T)$$

holds for all t less than T. Here, γ is some deterministic $\mathbb{R}^n$-valued function which is absolutely continuous with respect to T.

Then it follows that, under the martingale measure $\mathbb{Q}$, L obeys the SDE

$$d_t L(t,T) = L(t,T) \sum_{i=1}^{n} \gamma_i(t,T) \left(dW_i(t) + \left(\int_t^{T+\delta} \sigma_i(t,u) \, du \right) dt \right).$$

More interestingly, under the forward measure $\mathbb{P}_{T+\delta}$ (see section 6.4), L obeys

$$d_t L(t,T) = L(t,T) \sum_{i=1}^{n} \gamma_i(t,T) \, d\tilde{W}_i(t),$$

where $\tilde{W}_i$ are $\mathbb{P}_{T+\delta}$-Brownian motions. Thus $L(t, T)$, as a t-process, is not only a $\mathbb{P}_{T+\delta}$-martingale, but is also log-normal. We shall see later that this enables us to price caps and swaptions easily.

To price, we only need to know the function γ, rather than the whole volatility structure. While the γ function represents the correlation at time t between changes in the LIBOR rates at different forward dates T, in practice γ is calibrated by comparing the model's prices with the market. For instance, in their paper, Brace, Gatarek and Musiela fit a γ function of the form

$$\gamma_i(t, T) = f(t)\gamma_i(T - t)$$

by calibrating against known prices of caps and swaptions.

Writing $L(T)$ for $L(T, T)$, the instantaneous LIBOR rate, suppose we have a contract which pays off at a sequence of times $T_i = T_0 + i\delta$ $(i = 1, \ldots, n)$. If the payment at time T_{i+1} depends on the LIBOR rate set at time T_i, for example if $X = f(L(T_i))$, then the value of that payment at time t is

$$V_t = P(t, T) \mathbb{E}_{\mathbb{P}_{T_{i+1}}} \left(f(L(T_i)) \,\Big|\, \mathcal{F}_t \right).$$

The fact that $L(T_i)$ is log-normally distributed under $\mathbb{P}_{T_{i+1}}$ allows us to evaluate this expression for simple f.

One such simple f is the caplet payoff $\delta(L(T_{i-1}) - k)^+$ at time T_i. In this case, the worth of the caplet at time t is V_t, equal to

$$\delta P(t, T_i) \left\{ F\Phi\left(\frac{\log \frac{F}{k} + \frac{1}{2}\zeta^2(t, T_{i-1})}{\zeta(t, T_{i-1})}\right) - k\Phi\left(\frac{\log \frac{F}{k} - \frac{1}{2}\zeta^2(t, T_{i-1})}{\zeta(t, T_{i-1})}\right) \right\},$$

where F is the forward LIBOR rate $L(t, T_{i-1})$ and $\zeta^2(t, T)$ is $\int_t^T |\gamma(s, T)|^2 \, ds$, the variance of $\log L(T)$ given $\mathcal{F}_t$. This valuation has the familiar Black–Scholes form because under the forward measure $\mathbb{P}_{T_i}$, $L(T_{i-1})$ is log-normal and the calculation proceeds as usual.

We can even (approximately) price swaptions. Consider the option to pay fixed at rate k and receive floating and at times $T_i = T_0 + i\delta$ $(i = 1, \ldots, n)$. Let us set

$$\Gamma_i^2 = \int_t^{T_0} |\gamma(s, T_{i-1})|^2 \, ds,$$

which is the variance of $\log L(T_0, T_{i-1})$ given $\mathcal{F}_t$ under the forward measure $\mathbb{P}_{T_i}$. We also define

$$d_i = \sum_{j=1}^i \frac{\delta L(t, T_{j-1})}{1 + \delta L(t, T_{j-1})} \Gamma_j - \tfrac{1}{2}\Gamma_i,$$

and s_0 to be the unique root of the equation

$$s : \sum_{i=1}^{n} (k\delta + I(i = n)) \left\{ \prod_{j=1}^{i} \left(1 + \delta L(t, T_{j-1}) \exp(\Gamma_j(s + d_j)) \right) \right\}^{-1} = 1.$$

Then an approximation to the value at time t of the above swaption is

$$V_t = \delta \sum_{i=1}^{n} P(t, T_i) \left\{ L(t, T_{i-1}) \Phi\left(\frac{F_i + \frac{1}{2}\Gamma_i^2}{\Gamma_i} \right) - k\Phi\left(\frac{F_i - \frac{1}{2}\Gamma_i^2}{\Gamma_i} \right) \right\},$$

where $F_i = -\Gamma_i(s_0 + d_i)$.

Chapter 6
Bigger models

The Black–Scholes stock model assumes that the stock drift and stock volatility are constant. It assumes that there is only a single stock in the market. And it assumes that the cash bond is deterministic with zero volatility. None of these assumptions is necessary. The subsequent sections tackle these restrictions one by one and show how a more general model can still price and hedge derivatives. Also we will reveal the underlying framework which governs all these models from behind the scenes.

This is not to say that all models, no matter how complex or bizarre, will always give good prices. But if a model is driven by Brownian motions, and has no transaction costs, it is analysable in this framework.

6.1 General stock model

We recall that the Black–Scholes model contained a bond and a stock B_t and S_t with SDEs

$$dB_t = rB_t \, dt,$$

$$\text{and} \qquad dS_t = S_t(\sigma \, dW_t + \mu \, dt).$$

Here r is the constant interest rate, σ is the constant stock volatility and μ is the constant stock drift, and we are using the SDE formulation discussed in section 4.4. The process W is $\mathbb{P}$-Brownian motion.

Our most general stochastic process can have variable drift and volatility. Not only can they vary with time, but they can depend on movements of the stock itself (or equivalently, on movements of the Brownian motion W). We could replace the constant σ by a function of the stock price $\sigma(S_t)$, or even a function of both the stock price and time $\sigma(S_t, t)$. Even this is not fully general. (For instance the volatility at time t might depend on the maximum value achieved by the stock price up to time t.) We will replace σ by a general $\mathcal{F}$-previsible process σ_t, and the constants r and μ by $\mathcal{F}$-previsible processes r_t and μ_t respectively. The new SDEs are now

$$dB_t = r_t B_t \, dt,$$

$$\text{and} \qquad dS_t = S_t(\sigma_t \, dW_t + \mu_t \, dt).$$

These have solutions

$$B_t = \exp\left(\int_0^t r_s \, ds\right),$$

$$S_t = S_0 \exp\left(\int_0^t \sigma_s \, dW_s + \int_0^t (\mu_s - \tfrac{1}{2}\sigma_s^2) \, ds\right).$$

[Technical note: the processes σ_t, r_t and μ_t cannot be fully general, as they must be integrable enough for these integrals to exist. Explicitly, we need that (with $\mathbb{P}$-probability one), the integrals $\int_0^T \sigma_t^2 \, dt$, $\int_0^T |r_t| \, dt$, and $\int_0^T |\mu_t| \, dt$ are finite.]

Change of measure

As before, we aim to make the discounted stock price $Z_t = B_t^{-1} S_t$ into a martingale. This is achieved by adding a drift γ_t to W. That is, if $\tilde{W}_t = W_t + \int_0^t \gamma_s \, ds$ is $\mathbb{Q}$-Brownian motion, then Z_t has SDE

$$dZ_t = Z_t(\sigma_t \, d\tilde{W}_t + (\mu_t - r_t - \sigma_t \gamma_t) \, dt).$$

And Z is a $\mathbb{Q}$-martingale if

$$\gamma_t = \frac{\mu_t - r_t}{\sigma_t},$$

as was adumbrated in the market price of risk section (4.4). Now the market price of risk depends on the time t and the sample path up to that time. It will, however, continue to be independent of the instrument considered. It should also be checked, in any actual case, that γ_t satisfies the C-M-G growth condition $\mathbb{E}_{\mathbb{P}}(\exp \tfrac{1}{2} \int_0^T \gamma_t^2 \, dt) < \infty$.

Under $\mathbb{Q}$, Z has the SDE

$$dZ_t = \sigma_t Z_t \, d\tilde{W}_t,$$

so it is at least a local martingale because it is driftless. It should also be checked that Z is a proper martingale. For instance, it is enough that $\mathbb{E}_{\mathbb{Q}}(\exp \frac{1}{2} \int_0^T \sigma_t^2 \, dt)$ is finite.

Replicating strategies

If X is the derivative to be priced, with maturity at time T, then the procedure is not much different from the basic Black–Scholes technique.

We can form a $\mathbb{Q}$-martingale E_t through the conditional expectation process of the discounted claim, $E_t = \mathbb{E}_{\mathbb{Q}}(B_T^{-1} X | \mathcal{F}_t)$. Then the martingale representation theorem (section 3.5) says that the martingale E_t is the integral

$$E_t = E_0 + \int_0^t \phi_s \, dZ_s,$$

for some $\mathcal{F}$-previsible process ϕ_t. (Note that we need σ_t never to be zero.) Let us take ϕ_t to be our stock portfolio holding at time t. Then

$$dE_t = \phi_t \, dZ_t.$$

Setting the bond portfolio holding ψ_t to be $\psi_t = E_t - \phi_t Z_t$, then the value of the portfolio at time t is

$$V_t = \phi_t S_t + \psi_t B_t = B_t E_t.$$

It also follows (as in chapter three) that (ϕ, ψ) is self-financing in that the changes in the value V_t are due only to changes in the assets' prices. That is

$$dV_t = \phi_t \, dS_t + \psi_t \, dB_t.$$

So (ϕ, ψ) is a self-financing strategy with initial value $V_0 = \mathbb{E}_{\mathbb{Q}}(B_T^{-1} X)$ and terminal value $V_T = X$.

Derivative pricing

Arbitrage arguments convince us that the only value for the derivative at time t is

Derivative price

$$V_t = B_t \, \mathbb{E}_\mathbb{Q}\big(B_T^{-1}X \mid \mathcal{F}_t\big) = \mathbb{E}_\mathbb{Q}\Big(\exp\big(-\textstyle\int_t^T r_s \, ds\big)X \mid \mathcal{F}_t\Big).$$

In other words, the value at time t is the suitably discounted expectation of the derivative conditional on the history up to time t, under the measure which makes the discounted stock process a martingale – the risk-neutral measure.

There is no general expression which will provide a more explicit answer for the option value V_t. To make specific calculations, one needs to know the discount rate r_t, the volatility of the stock σ_t – though not its drift – and the derivative itself.

Implementation

In practice, if the model is much more complex than Black–Scholes, these expectations cannot be performed analytically. (The log-normal cases of section 6.2 will be notable exceptions.) Instead numerical methods must be used.

If we can approximate the price V_t at time t, then an approximation for ϕ_t or "dV_t/dS_t" is the *delta hedge*

$$\phi_t \approx \frac{\Delta V_t}{\Delta S_t},$$

where Δ represents the change over a small time interval $(t, t + \Delta t)$.

6.2 Log-normal models

We have already seen that the Black–Scholes formula can be true, even if we are not working with the Black–Scholes model (as in section 4.1). The common feature of models where this happens is that the asset prices are log-normally distributed under the martingale measure $\mathbb{Q}$.

In the simple Black–Scholes model, the cash bond and the stock are modelled as

$$B_t = e^{rt} \quad \text{and} \quad S_t = S_0 \exp(\sigma W_t + \mu t),$$

where r, σ and μ are constants and W is $\mathbb{P}$-Brownian motion. The forward price to purchase F at time T is

$$F = S_0 e^{rT}.$$

And the value at time zero of an option to buy S_T for a strike price of k is

$$V_0 = e^{-rT} \left\{ F\Phi\left(\frac{\log \frac{F}{k} + \frac{1}{2}\sigma^2 T}{\sigma\sqrt{T}} \right) - k\Phi\left(\frac{\log \frac{F}{k} - \frac{1}{2}\sigma^2 T}{\sigma\sqrt{T}} \right) \right\}.$$

Log-normal asset prices

When prices, under the martingale measure, are log-normal, there are great advantages. This holds for the Black–Scholes model itself, for some currency and equity models, and also for simple interest rate models.

Explicitly, suppose the stock S_T and the cash bond B_T are known to be jointly log-normally distributed under the martingale measure $\mathbb{Q}$. Let $\sigma_1^2 T$ be the variance of $\log S_T$, $\sigma_2^2 T$ be the variance of $\log B_T^{-1}$ (σ_1 and σ_2 are term volatilities), and let ρ be their correlation. Then the forward price for purchasing S at time T is

$$F = \frac{\mathbb{E}_\mathbb{Q}(B_T^{-1} S_T)}{\mathbb{E}_\mathbb{Q}(B_T^{-1})}, \quad \text{or equivalently} \quad F = \exp(\rho\sigma_1\sigma_2 T)\mathbb{E}_\mathbb{Q}(S_T),$$

and the price of a call on S_T struck at k is the generalised Black–Scholes formula

$$V_0 = \mathbb{E}_\mathbb{Q}(B_T^{-1}) \left\{ F\Phi\left(\frac{\log \frac{F}{k} + \frac{1}{2}\sigma_1^2 T}{\sigma_1\sqrt{T}} \right) - k\Phi\left(\frac{\log \frac{F}{k} - \frac{1}{2}\sigma_1^2 T}{\sigma_1\sqrt{T}} \right) \right\}.$$

We can see why these formulae are true. Write S_T as

$$S_T = A \exp\left(\alpha_1 Z - \tfrac{1}{2}\alpha_1^2\right), \quad \text{with } \alpha_1^2 = \sigma_1^2 T,$$

where A is the constant $\mathbb{E}_\mathbb{Q}(S_T)$ and Z is a normal $N(0,1)$ random variable under $\mathbb{Q}$. The discount factor B_T^{-1} is log-normal with log-variance $\sigma_2^2 T$ and its correlation with the stock log-price is ρ. Setting B to be its expectation $B = \mathbb{E}_{\mathbb{Q}}(B_T^{-1})$, we get

$$B_T^{-1} = B \exp\left(\alpha_2(\rho Z + \bar{\rho}W) - \tfrac{1}{2}\alpha_2^2\right), \qquad \text{with } \alpha_2^2 = \sigma_2^2 T,$$

where $\bar{\rho} = \sqrt{1 - \rho^2}$ and W is a normal $N(0, 1)$ independent of Z.

The expected discounted stock price is then

$$\mathbb{E}_{\mathbb{Q}}(B_T^{-1} S_T) = AB \exp\left(\tfrac{1}{2}(\alpha_1 + \rho\alpha_2)^2 + \tfrac{1}{2}\bar{\rho}^2\alpha_2^2 - \tfrac{1}{2}\alpha_1^2 - \tfrac{1}{2}\alpha_2^2\right) = AB \exp(\rho\alpha_1\alpha_2).$$

So the forward price for S_T is thus $F = A \exp(\rho\sigma_1\sigma_2 T)$. Re-expressing S_T:

$$S_T = F \exp(\alpha_1 Z - \tfrac{1}{2}\alpha_1^2 - \rho\alpha_1\alpha_2),$$

gives us the call value

$$V_0 = \mathbb{E}_{\mathbb{Q}}\left(B_T^{-1}(S_T - k)^+\right) = B\,\mathbb{E}_{\mathbb{Q}}\left(e^{\rho\alpha_2 Z - \tfrac{1}{2}\rho^2\alpha_2^2}(S_T - k)^+\right),$$

which is also equal to

$$B\,\mathbb{E}_{\mathbb{Q}}\left(Fe^{(\alpha_1 + \rho\alpha_2)Z - \tfrac{1}{2}(\alpha_1 + \rho\alpha_2)^2} - ke^{\rho\alpha_2 Z - \tfrac{1}{2}\rho^2\alpha_2^2}\,; Z \geqslant -z\right),$$

where z is the critical value $z = (\log \frac{F}{k} - \tfrac{1}{2}\alpha_1^2 - \rho\alpha_1\alpha_2)/\alpha_1$. Using the probabilistic result that $\mathbb{E}\left(e^{yZ - \tfrac{1}{2}y^2}\,; Z \geqslant -z\right) = \Phi(y + z)$, for any constants y and z, the result follows. [The notation $\mathbb{E}(X; A)$ denotes the expectation of the random variable X over the event A, or equivalently is $\mathbb{E}(XI_A)$, where I_A is the indicator function of the event A.]

6.3 Multiple stock models

Black–Scholes assumes a single stock in the market. In many cases, this assumption does little harm. If we write an option on, say, General Motors stock, having modelled its behaviour adequately, we are unaffected by the movements of other securities. However, more complex equity products, such as quantos, depend on the behaviour of at least two separate securities. Even more so in the bond market, where a swap's current value is affected by the movements of a large number of bonds of varying maturities.

A good model of several securities must not only describe each one individually, but also represent the interaction and dependency between them. For instance, our quanto contract of section 4.5 was related to both the sterling/dollar exchange rate and an individual UK stock. These two processes have some degree of co-dependence. In particular, large movements in one may be linked with corresponding movements in the other. Such changes would suggest that the two securities are correlated.

Stochastic processes adapted to n-dimensional Brownian motion

A stochastic process X is a continuous process $(X_t : t \geqslant 0)$ such that X_t can be written as

$$X_t = X_0 + \sum_{i=1}^{n} \int_0^t \sigma_i(s) \, dW_s^i + \int_0^t \mu_s \, ds,$$

where $\sigma_1, \ldots, \sigma_n$ and μ are random $\mathcal{F}$-previsible processes such that the integral $\int_0^t (\sum_i \sigma_i^2(s) + |\mu_s|) \, ds$ is finite for all times t (with probability 1). The differential form of this equation can be written

$$dX_t = \sum_{i=1}^{n} \sigma_i(t) \, dW_t^i + \mu_t \, dt.$$

Multiple stocks can be driven by multiple Brownian motions. Instead of just one $\mathbb{P}$-Brownian motion, we will have, in the n-factor case, n independent Brownian motions $W_t^1, \ldots, W_t^n$. That means that each W_t^i behaves as a Brownian motion, and the behaviour of any one of them is completely uninfluenced by the movements of the others. Their filtration $\mathcal{F}_t$ is now the total of all the histories of the n Brownian motions. In other words, $\mathcal{F}_T$ is the history of the n-dimensional vector $(W_t^1, \ldots, W_t^n)$ up to time T. This leads to an enhanced definition of a stochastic process (see box).

The drift term is unchanged from the original (one-factor) definition, but there is now a volatility process $\sigma_i(t)$ for each factor. We must remember that in a multi-factor setting volatility is no longer a scalar, but strictly is now a vector. The total volatility of the process X is $\sqrt{\sigma_1^2(t) + \ldots + \sigma_n^2(t)}$. In other words, the variance of dX_t is $\sum_i \sigma_i^2(t) \, dt$, made up of the contribution $\sigma_i^2(t) \, dt$ from each Brownian motion component W^i, the variances adding because the Brownian motion components are independent.

There is also an n–factor version of Itô's formula and the product rule.

> **Itô's formula (n-factor)**
>
> If X is a stochastic process, satisfying $dX_t = \sum_i \sigma_i(t) \, dW_t^i + \mu_t \, dt$, and f is a deterministic twice continuously differentiable function, then $Y_t := f(X_t)$ is also a stochastic process with stochastic increment
>
> $$dY_t = \sum_{i=1}^{n} \Big(\sigma_i(t) f'(X_t) \Big) \, dW_t^i + \Big(\mu_t f'(X_t) + \tfrac{1}{2} \sum_{i=1}^{n} \sigma_i^2(t) f''(X_t) \Big) \, dt.$$

Again this is an analogue of the one-factor Itô formula, with the replication of the volatility terms for each additional Brownian factor.

> **Product rule (n-factor)**
>
> If X is a stochastic process satisfying $dX_t = \sum_i \sigma_i(t) \, dW_t^i + \mu_t \, dt$, and Y is a stochastic process satisfying $dY_t = \sum_i \rho_i(t) \, dW_t^i + \nu_t \, dt$, then $X_t Y_t$ is a stochastic process satisfying
>
> $$d(X_t Y_t) = X_t \, dY_t + Y_t \, dX_t + \Big(\sum_{i=1}^{n} \sigma_i(t) \rho_i(t) \Big) \, dt.$$

This new version unifies the two apparently different cases of the product rule we encountered in section 3.3. If X_t and Y_t are both adapted to the same Brownian motion W_t, then this rule agrees with the first case. If however X_t and Y_t are adapted to two independent Brownian motions, say W_t^1 and W_t^2, then X_t will have zero volatility with respect to W^2, that is $\sigma_2(t) = 0$, and similarly Y_t will have zero volatility with respect to W^1, $\rho_1(t) = 0$. Thus the term $\sum \sigma_i(t) \rho_i(t)$ in the n-factor product rule will be identically zero, agreeing with the second case in section 3.3.

The Cameron–Martin–Girsanov theorem continues to hold where W is n-dimensional Brownian motion and the drift γ is an n-vector process for which $\mathbb{E}_{\mathbb{P}} \exp(\tfrac{1}{2} \int_0^T |\gamma_t|^2 \, dt)$ is finite.

> **Cameron–Martin–Girsanov theorem (n-factor)**
>
> Let $W = (W^1, \ldots, W^n)$ be n-dimensional $\mathbb{P}$-Brownian motion. Suppose that $\gamma_t = (\gamma_t^1, \ldots, \gamma_t^n)$ is an $\mathcal{F}$-previsible n-vector process which satisfies the growth condition $\mathbb{E}_{\mathbb{P}} \exp\left(\frac{1}{2}\int_0^T |\gamma_t|^2 \, dt\right) < \infty$, and we set $\tilde{W}_t^i = W_t^i + \int_0^t \gamma_s^i \, ds$. Then there is a new measure $\mathbb{Q}$, equivalent to $\mathbb{P}$ up to time T, such that $\tilde{W} := (\tilde{W}^1, \ldots, \tilde{W}^n)$ is n-dimensional $\mathbb{Q}$-Brownian motion up to time T.
>
> The Radon–Nikodym derivative of $\mathbb{Q}$ by $\mathbb{P}$ is
>
> $$\frac{d\mathbb{Q}}{d\mathbb{P}} = \exp\left(-\sum_{i=1}^n \int_0^T \gamma_t^i \, dW_t^i - \frac{1}{2}\int_0^T |\gamma_t|^2 \, dt\right).$$

There is also a converse to this theorem, exactly analagous to the one-factor converse.

Finally, we recall from section 5.5 that there is an n-factor martingale representation theorem. With W as n-dimensional $\mathbb{Q}$-Brownian motion, M as an n-dimensional $\mathbb{Q}$-martingale with non-singular volatility matrix, and N any other one-dimensional $\mathbb{Q}$-martingale, then there is an $\mathcal{F}$-previsible n-vector process $\phi_t = (\phi_t^1, \ldots, \phi_t^n)$ such that

$$N_t = N_0 + \sum_{j=1}^n \int_0^t \phi_s^j \, dM_s^j.$$

The general n-factor model

We will see later that it is important that we have essentially as many basic securities (excluding the cash bond) as there are Brownian factors. Generally speaking, if there are more securities than factors there might be arbitrage, and if there are fewer we will not be able to hedge. The situation is not quite as simple as that (the bond market, for instance, has an unlimited number of different maturity bonds), but we shall start with the canonical case.

Our model then, will contain a cash bond B_t as usual, and n different market securities $S_t^1, \ldots, S_t^n$. Their SDEs are

$$dB_t = r_t B_t \, dt,$$

$$dS_t^i = S_t^i \left(\sum_{j=1}^n \sigma_{ij}(t) \, dW_t^j + \mu_t^i \, dt\right), \quad i = 1, \ldots, n.$$

Here r_t is the instantaneous short-rate process, μ_t^i is the drift of the ith security, and $(\sigma_{ij})_{j=1}^n$ is its volatility vector. As each security has a volatility vector, the collection of n such vectors forms a volatility matrix $\Sigma_t = \left(\sigma_{ij}(t)\right)_{i,j=1}^n$ of processes. In integral form, these securities are

$$B_t = \exp\left(\int_0^t r_s\, ds\right),$$

$$S_t^i = S_0^i \exp\left(\sum_{j=1}^n \int_0^t \sigma_{ij}(s)\, dW_s^j + \int_0^t \left(\mu_s^i - \tfrac{1}{2}\sum_{j=1}^n \sigma_{ij}^2(s)\right) ds\right).$$

Change of measure

We now want to find a new measure $\mathbb{Q}$, under which *all* the discounted stock prices are $\mathbb{Q}$-martingales simultaneously.

Suppose we add a drift $\gamma_t = (\gamma_t^1, \ldots, \gamma_t^n)$ to W_t, so that

$$\tilde{W}_t^i = W_t^i + \int_0^t \gamma_s^i\, ds$$

is $\mathbb{Q}$-Brownian motion, by the n-factor C–M–G theorem. Then the discounted stock price $Z_t^i = B_t^{-1} S_t^i$ has SDE

$$dZ_t^i = Z_t^i \left(\sum_{j=1}^n \sigma_{ij}(t)\, d\tilde{W}_t^j + \left(\mu_t^i - r_t - \sum_{j=1}^n \sigma_{ij}(t)\gamma_t^j\right) dt\right).$$

To make the drift term vanish for each i, we must have that

$$\sum_{j=1}^n \sigma_{ij}(t)\gamma_t^j = \mu_t^i - r_t, \qquad \text{for all } t, \ i = 1, \ldots, n.$$

In terms of vectors and matrices, this can be re-expressed as

$$\Sigma_t \gamma_t = \mu_t - r_t \mathbf{1},$$

where Σ_t is the matrix $(\sigma_{ij}(t))$ and $\mathbf{1}$ is the constant vector $(1, 1, \ldots, 1)$. This vector equation may or may not have a solution γ_t for any particular t. Whether it does or not depends on the actual values of Σ_t, μ_t and r_t. If, though, the matrix Σ_t is invertible, then a unique such γ_t must exist and be equal to

$$\gamma_t = \Sigma_t^{-1}(\mu_t - r_t \mathbf{1}).$$

The one–factor market price of risk formula $\gamma_t = \sigma_t^{-1}(\mu_t - r_t)$ is now just a special case. This means that if Σ_t is invertible for every t and γ_t satisfies the C–M–G condition $\mathbb{E}_{\mathbb{P}} \exp(\frac{1}{2}\int_0^T |\gamma_t|^2 \, dt) < \infty$, then there is a measure $\mathbb{Q}$ which makes the discounted stock prices into $\mathbb{Q}$-martingales. (Or at least into $\mathbb{Q}$-local martingales. We also need the integral condition that for each i, $\mathbb{E}_{\mathbb{Q}}(\exp \frac{1}{2}\sum_j^n \int_0^T \sigma_{ij}^2(t) \, dt) < \infty$, for Z^i to be a proper $\mathbb{Q}$-martingale.)

Replicating strategies

Let X be a derivative maturing at time T, and let E_t be the $\mathbb{Q}$-martingale $E_t = \mathbb{E}_{\mathbb{Q}}(B_T^{-1}X|\mathcal{F}_t)$. If the matrix Σ_t is always invertible, then the n-factor martingale representation theorem gives us a volatility vector process $\phi_t = (\phi_t^1, \ldots, \phi_t^n)$ such that

$$E_t - E_0 + \sum_{j=1}^n \int_0^t \phi_s^j \, dZ_s^j.$$

The invertibility of Σ_t is essential at this stage. Our hedging strategy will be $(\phi_t^1, \ldots, \phi_t^n, \psi_t)$ where ϕ_t^i is the holding of security i at time t and ψ_t is the bond holding. As usual, the bond holding ψ is

$$\psi_t = E_t - \sum_{j=1}^n \phi_t^j Z_t^j,$$

so that the value of the portfolio is $V_t = B_t E_t$. The portfolio is self-financing, in that

$$dV_t = \sum_{j=1}^n \phi_t^j \, dS_t^j + \psi_t \, dB_t.$$

Derivative pricing

The value of the derivative at time t is

$$V_t = B_t \, \mathbb{E}_{\mathbb{Q}}\left(B_T^{-1}X \mid \mathcal{F}_t\right) = \mathbb{E}_{\mathbb{Q}}\left(\exp\left(-\int_t^T r_s \, ds\right)X \mid \mathcal{F}_t\right).$$

6.4 Numeraires

Although the numeraire is usually chosen to be a cash bond, it needn't be. In fact, not only can the numeraire have volatility, it can be any of the tradable instruments available. We have seen in the foreign exchange context that there can be a choice of which currency's cash bond to use. But no matter which numeraire is chosen, the price of the derivative will always be the same. It is because the choice of numeraire doesn't matter, that we usually pick the stolid cash bond.

When we proved the self-financing condition in chapter three, we assumed that the numeraire had no volatility. This is not actually necessary. But we do have to check that the self-financing equations will still work. We want to show that

Self-financing strategies

A portfolio strategy (ϕ_t, ψ_t) of holdings in a stock S_t and a possibly volatile cash bond B_t has value $V_t = \phi_t S_t + \psi_t B_t$ and discounted value $E_t = \phi_t Z_t + \psi_t$, where Z is the discounted stock process $Z_t = B_t^{-1} S_t$. Then the strategy is self-financing if either

$$dV_t = \phi_t\, dS_t + \psi_t\, dB_t, \quad \text{or equivalently} \quad dE_t = \phi_t\, dZ_t.$$

Recall the one-factor product rule

$$d(XY)_t = X_t\, dY_t + Y_t\, dX_t + \sigma_t \rho_t\, dt,$$

where X and Y are stochastic processes with stochastic differentials

$$dX_t = \sigma_t\, dW_t + \mu_t\, dt,$$
$$dY_t = \rho_t\, dW_t + \nu_t\, dt.$$

Suppose we have a strategy (ϕ, ψ), with discounted value E_t satisfying $dE_t = \phi_t\, dZ_t$. We want to show that (ϕ, ψ) is self-financing. We do this with two applications of the product rule. Firstly

$$dV_t = d(B_t E_t) = B_t\, dE_t + E_t\, dB_t + \sigma_t(\phi_t \rho_t)\, dt,$$

where σ_t is the volatility of B_t and ρ_t is the volatility of Z_t (and hence $\phi_t \rho_t$ is the volatility of E_t). We can use the substitutions $dE_t = \phi_t \, dZ_t$ and $E_t = \phi_t Z_t + \psi_t$ to rearrange the above expression into

$$dV_t = \phi_t (B_t \, dZ_t + Z_t \, dB_t + \sigma_t \rho_t \, dt) + \psi_t \, dB_t.$$

The second use of the product rule says that the term in brackets above is equal to $d(BZ)_t = dS_t$. The resulting equation is the self-financing equation. This also holds for n-factor models with multiple stocks.

Changing numeraires

Suppose we have a number of securities including some stocks $S_t^1, \ldots, S_t^n$ and two others B_t and C_t either of which might be a numeraire. If we choose B_t to be our numeraire, we need to find a measure $\mathbb{Q}$ (equivalent to the original measure) under which

$$B_t^{-1} S_t^i \quad (i = 1, \ldots, n) \qquad \text{and} \qquad B_t^{-1} C_t$$

are $\mathbb{Q}$-martingales. Then the value at time t of a derivative payoff X at time T is

$$V_t = B_t \, \mathbb{E}_{\mathbb{Q}} (B_T^{-1} X \mid \mathcal{F}_t).$$

Suppose however that we choose C_t to be our numeraire instead. Then we would have a different measure $\mathbb{Q}^C$ under which

$$C_t^{-1} S_t^i \quad (i = 1, \ldots, n) \qquad \text{and} \qquad C_t^{-1} B_t$$

are $\mathbb{Q}^C$-martingales. We can actually find out what $\mathbb{Q}^C$ is, or at least what its Radon–Nikodym derivative with respect to $\mathbb{Q}$ is. We recall Radon–Nikodym fact (ii) from section 3.4, that for any process X_t,

$$\zeta_s \, \mathbb{E}_{\mathbb{Q}^C} (X_t \mid \mathcal{F}_s) = \mathbb{E}_{\mathbb{Q}} (\zeta_t X_t \mid \mathcal{F}_s),$$

where ζ_t is the change of measure process $\zeta_t = \mathbb{E}_{\mathbb{Q}} (\frac{d\mathbb{Q}^C}{d\mathbb{Q}} | \mathcal{F}_t)$. It follows from this that if X_t happens to be a $\mathbb{Q}^C$-martingale, then

$$\zeta_s X_s = \mathbb{E}_{\mathbb{Q}} (\zeta_t X_t \mid \mathcal{F}_s),$$

and so $\zeta_t X_t$ is a $\mathbb{Q}$-martingale.

The canonical $\mathbb{Q}^C$-martingales (including the constant martingale with value 1) are 1, $C_t^{-1}B_t$, $C_t^{-1}S_t^1$, ..., $C_t^{-1}S_t^n$ and similarly the $\mathbb{Q}$-martingales are $B_t^{-1}C_t$, 1, $B_t^{-1}S_t^1$, ..., $B_t^{-1}S_t^n$. Each corresponding pair has a common ratio of $\zeta_t = B_t^{-1}C_t$. Thus the Radon–Nikodym derivative of $\mathbb{Q}^C$ with respect to $\mathbb{Q}$ is the ratio of the numeraire C to the numeraire B,

$$\frac{d\mathbb{Q}^C}{d\mathbb{Q}} = \frac{C_T}{B_T}.$$

The price of a payoff X maturing at T under the $\mathbb{Q}^C$ measure is

$$V_t^C = C_t\,\mathbb{E}_{\mathbb{Q}^C}\left(C_T^{-1}X \mid \mathcal{F}_t\right).$$

Using again the Radon–Nikodym result that $\mathbb{E}_{\mathbb{Q}^C}(X|\mathcal{F}_t) = \zeta_t^{-1}\mathbb{E}_{\mathbb{Q}}(\zeta_T X|\mathcal{F}_t)$, then

$$V_t^C = \zeta_t^{-1}C_t\,\mathbb{E}_{\mathbb{Q}}\left(\zeta_T C_T^{-1}X \mid \mathcal{F}_t\right) = B_t\,\mathbb{E}_{\mathbb{Q}}\left(B_T^{-1}X \mid \mathcal{F}_t\right).$$

This is exactly the same as the price V_t under $\mathbb{Q}$, so the two agree, just as in the foreign exchange section (4.1), where the dollar and sterling investors agreed on all derivative prices.

Example – forward measures in the interest-rate market

In interest-rate models, it is often popular to use a bond maturing at date T (the T-bond with price $P(t,T)$) as the numeraire. The martingale measure for this numeraire is called the T-forward measure $\mathbb{P}_T$ and makes the forward rate $f(t,T)$ a $\mathbb{P}_T$-martingale, as well as the δ-period LIBOR rate for borrowing up till time T.

The new numeraire is the T-bond normalised to have unit value at time zero. If we call this numeraire C_t, then $C_t = P(t,T)/P(0,T)$. The forward measure $\mathbb{P}_T$ thus has Radon–Nikodym derivative with respect to $\mathbb{Q}$ of

$$\frac{d\mathbb{P}_T}{d\mathbb{Q}} = \frac{C_T}{B_T} = \frac{1}{P(0,T)B_T}.$$

The associated $\mathbb{Q}$-martingale is

$$\zeta_t = \mathbb{E}_{\mathbb{Q}}\left(\frac{d\mathbb{P}_T}{d\mathbb{Q}} \mid \mathcal{F}_t\right) = \frac{C_t}{B_t} = \frac{P(t,T)}{P(0,T)B_t}.$$

Now the forward price set at time t for purchasing X at date T is its current value V_t scaled up by the return on a T-bond, namely $F_t =$

$P^{-1}(t,T)B_t \, \mathbb{E}_{\mathbb{Q}}(B_T^{-1}X \mid \mathcal{F}_t)$. Once more, by property (ii) of the Radon–Nikodym derivative, F_t equals

$$F_t = \mathbb{E}_{\mathbb{P}_T}(X|\mathcal{F}_t),$$

so is itself a $\mathbb{P}_T$-martingale. Calculating the forward price for X is now only a matter of taking its expectation under the forward measure.

From the SDE for $P(t,T)$, we find that ζ_t satisfies

$$d\zeta_t = \zeta_t \sum_{i=1}^{n} \Sigma_i(t,T) \, dW_i(t),$$

where W is n-dimensional $\mathbb{Q}$-Brownian motion, and $\Sigma_i(t,T)$ is the component of the volatility of $P(t,T)$ with respect to $W_i(t)$. By the converse of the C–M–G theorem, we see that

$$\tilde{W}_i(t) = W_i(t) - \int_0^t \Sigma_i(s,T) \, ds$$

is $\mathbb{P}_T$-Brownian motion.

This gives an alternative expression for pricing interest-rate derivatives. If X is a payoff at date T, then its value at time t is

$$V_t = B_t \, \mathbb{E}_{\mathbb{Q}}(B_T^{-1}X \mid \mathcal{F}_t) = P(t,T) \, \mathbb{E}_{\mathbb{P}_T}(X|\mathcal{F}_t).$$

So the value of X at time t is just the $\mathbb{P}_T$-expectation of X up to time t (the forward price of X) discounted by the (T-bond) time value of money up to date T.

Also the forward rates $f(t,T)$ are the forward rates for r_T, so that $f(t,T)$ is a $\mathbb{P}_T$-martingale with

$$f(t,T) = \mathbb{E}_{\mathbb{P}_T}(r_T|\mathcal{F}_t),$$

and $\qquad d_t f(t,T) = \displaystyle\sum_{i=1}^{n} \sigma_i(t,T) \, d\tilde{W}_i(t).$

Another forward measure martingale is the δ-period LIBOR rate

$$L_t = \frac{1}{\delta}\left(\frac{P(t,T-\delta)}{P(t,T)} - 1\right).$$

See chapter five (section 5.7) for more details.

6.5 Foreign currency interest-rate models

We have looked at foreign exchange (section 4.1). We have looked at the interest rate market (chapter five). But we have not yet studied an interest rate market of another currency. Now we will.

For definiteness, we will imagine ourselves to be a dollar investor operating in both the dollar and sterling interest-rate markets. Our variables will be

Table 6.1 Notation

$P(t,T)$: the dollar zero–coupon bond market prices
$f(t,T)$: the forward rate of dollar borrowing at date T (is $-\frac{\partial}{\partial T}\log P(t,T)$)
$\sigma(t,T)$: the volatility of $f(t,T)$
$\alpha(t,T)$: the drift of $f(t,T)$
  r_t : the dollar short rate (equal to $f(t,t)$)
  B_t : the dollar cash bond (equal to $\exp\int_0^t r_s\,ds$)

$Q(t,T)$: the sterling zero–coupon bond market prices
$g(t,T)$: the forward rate of sterling borrowing at date T (is $-\frac{\partial}{\partial T}\log Q(t,T)$)
$\tau(t,T)$: the volatility of $g(t,T)$
$\beta(t,T)$: the drift of $g(t,T)$
  u_t : the sterling short rate (equal to $g(t,t)$)
  D_t : the sterling cash bond (equal to $\exp\int_0^t u_s\,ds$)

  C_t : the exchange rate value in dollars of one pound
  ρ_t : the log-volatility of the exchange rate
  λ_t : the drift coefficient of the exchange rate (the drift of dC_t/C_t).

As in the HJM model, we will work in an n-factor model driven by the independent Brownian motions $W_t^1,\dots,W_t^n$. Of course n might be one, but it needn't be, in which case, the volatilities σ, τ and ρ are n-vectors $\sigma_i(t,T)$, $\tau_i(t,T)$ and $\rho_i(t)$ $(i=1,\dots,n)$.

What we have here are two separate interest-rate markets (the dollar denominated and the sterling denominated), plus a currency market linking them. The multi-factor model approach is needed to reflect varying degrees of correlation between various securities in the three markets.

The differentials of these processes are

$$d_t f(t,T) = \sum_{i=1}^{n} \sigma_i(t,T)\, dW_t^i + \alpha(t,T)\, dt,$$

$$d_t g(t,T) = \sum_{i=1}^{n} \tau_i(t,T)\, dW_t^i + \beta(t,T)\, dt,$$

$$dC_t = C_t \left(\sum_{i=1}^{n} \rho_i(t,T)\, dW_t^i + \lambda_t\, dt \right).$$

Apart from the dollar cash bond B_t, the dollar tradable securities in this market consist of the dollar-bonds $P(t,T)$; the dollar worth of the sterling bonds $C_t Q(t,T)$; and the dollar worth of the sterling cash bond $C_t D_t$. Let us fix T, and let the dollar discounted value of these three securities be X, Y and Z respectively, where

$$X_t = B_t^{-1} P(t,T),$$
$$Y_t = B_t^{-1} C_t Q(t,T),$$
$$Z_t = B_t^{-1} C_t D_t.$$

It will simplify later expressions to introduce the notation Σ_i, T_i and $\tilde{T}_i$, where

$$\Sigma_i(t,T) = -\int_t^T \sigma_i(t,u)\, du,$$

$$T_i(t,T) = -\int_t^T \tau_i(t,u)\, du,$$

$$\tilde{T}_i(t,T) = T_i(t,T) + \rho_i(t).$$

Then $\Sigma_i(t,T)$ is the W^i-volatility term of $P(t,T)$, $T_i(t,T)$ is the same for $Q(t,T)$, and $\tilde{T}_i(t,T)$ is the same for $C_t Q(t,T)$.

Our plan, much as ever, is to follow the three steps to replication. The first thing to do is to find a change of measure under which X_t, Y_t and Z_t are all martingales.

For any previsible n-vector $\gamma = (\gamma_i(t))_{i=1}^n$, there is a new measure $\mathbb{Q}$ and a $\mathbb{Q}$-Brownian motion $\tilde{W} = (\tilde{W}_t^1, \ldots, \tilde{W}_t^n)$, where $\tilde{W}_t^i = W_t^i + \int_0^t \gamma_i(s)\, ds$.

6.5 Foreign currency interest-rate models

Then the SDEs of X, Y and Z with respect to $\mathbb{Q}$ are

$$dX_t = X_t \left(\sum_{i=1}^{n} \Sigma_i(t,T) \, d\tilde{W}_t^i + \left(\int_t^T (\xi(t,u) - \alpha(t,u)) \, du \right) dt \right)$$

$$dY_t = Y_t \left(\sum_{i=1}^{n} \tilde{T}_i(t,T) \, d\tilde{W}_t^i + \left(\nu_t + \int_t^T (\eta(t,T) - \beta(t,u)) \, du \right) dt \right)$$

$$dZ_t = Z_t \left(\sum_{i=1}^{n} \rho_i(t) \, d\tilde{W}_t^i + \nu_t \, dt \right),$$

where $\xi(t,T)$, $\eta(t,T)$ and ν_t are defined to be

$$\xi(t,T) = \sum_{i=1}^{n} \sigma_i(t,u)\big(\gamma_i(t) - \Sigma_i(t,u)\big),$$

$$\eta(t,T) = \sum_{i=1}^{n} \tau_i(t,u)\big(\gamma_i(t) - \tilde{T}_i(t,u)\big),$$

$$\nu_t = \lambda_t - r_t + u_t - \sum_i \rho_i(t)\gamma_i(t).$$

Then there will be a martingale measure only if there is some choice of γ which makes all of X, Y and Z driftless. This happens if

$$\alpha(t,T) = \sum_{i=1}^{n} \sigma_i(t,T)\big(\gamma_i(t) - \Sigma_i(t,T)\big),$$

$$\beta(t,T) = \sum_{i=1}^{n} \tau_i(t,T)\big(\gamma_i(t) - \tilde{T}_i(t,T)\big),$$

$$\lambda_t = r_t - u_t + \sum_{i=1}^{n} \rho_i(t)\gamma_i(t).$$

Then under this $\mathbb{Q}$ measure

$$d_t P(t,T) = P(t,T) \left(\sum_{i=1}^{n} \Sigma_i(t,T) \, d\tilde{W}_t^i + r_t \, dt \right),$$

$$d_t Q(t,T) = Q(t,T) \left(\sum_{i=1}^{n} T_i(t,T) \, d\tilde{W}_t^i + \left(u_t - \sum_{i=1}^{n} \rho_i(t) T_i(t,T) \right) dt \right),$$

$$dC_t = C_t \left(\sum_{i=1}^{n} \rho_i(t) \, d\tilde{W}_t^i + (r_t - u_t) \, dt \right).$$

As long as this measure $\mathbb{Q}$ is unique, we will be able to hedge. (And uniqueness will follow if the volatility vectors of any n of the dollar tradable securities make an invertible matrix.) A derivative X paid in dollars at date T will have value at time t

$$V_t = B_t \, \mathbb{E}_{\mathbb{Q}}\big(B_T^{-1} X \mid \mathcal{F}_t\big).$$

The sterling investor

The sterling investor is on the other side of the mirror. He works with a different martingale measure $\mathbb{Q}^{\mathcal{L}}$. This reflects that his numeraire is the sterling cash bond D_t rather than the dollar cash bond. The Radon–Nikodym derivative of $\mathbb{Q}^{\mathcal{L}}$ with respect to $\mathbb{Q}$ will be the ratio of the dollar worth of the sterling bond to the dollar numeraire. (Normalising $D_0 = 1/C_0$ for convenience.) That is

$$\mathbb{E}_{\mathbb{Q}}\left(\frac{d\mathbb{Q}^{\mathcal{L}}}{d\mathbb{Q}} \,\bigg|\, \mathcal{F}_t\right) = \frac{C_t D_t}{B_t} = Z_t.$$

As Z_t has the SDE $dZ_t = Z_t \sum_i \rho_i(t)\, d\tilde{W}_t^i$, the difference in drifts between the $\mathbb{Q}^{\mathcal{L}}$-Brownian motion $\tilde{W}^{\mathcal{L}}$ and the $\mathbb{Q}$-Brownian motion $\tilde{W}$ is just ρ. That is

$$\tilde{W}_i^{\mathcal{L}}(t) = \tilde{W}_t^i - \int_0^t \rho_i(s)\, ds.$$

To the sterling investor, the sterling bonds have SDE

$$d_t Q(t,T) = Q(t,T) \left(\sum_{i=1}^n T_i(t,T)\, d\tilde{W}_i^{\mathcal{L}}(t) + u_t\, dt \right),$$

which is exactly the form that HJM leads us to expect.

As explained in section 6.4, the sterling investor will agree with the dollar investor on prices of future payoffs.

6.6 Arbitrage-free complete models

Time and again we have seen the same basic techniques used to price and hedge derivatives. Firstly, the C–M–G theorem is used to make the discounted price processes into martingales under a new measure. Then the martingale representation theorem gives a hedge for the derivative. The repeated recurrence of this program suggests that there might be a more general result underpinning it. And there is.

Before stating this canonical theorem, it is worth carefully laying out some concepts we have already brushed up against.

- **arbitrage-free.** A market is arbitrage-free if there is no way of making riskless profits. An arbitrage opportunity would be a (self-financing) trading strategy which started with zero value and terminated at some definite date T with a positive value. A market is arbitrage-free if there are absolutely no such arbitrage opportunities.

- **complete.** A market is said to be complete if any possible derivative claim can be hedged by trading with a self-financing portfolio of securities.

- **equivalent martingale measure (EMM).** Suppose we have a market of securities and a numeraire cash bond under a measure $\mathbb{P}$. An EMM is a measure $\mathbb{Q}$ equivalent to $\mathbb{P}$, under which the bond-discounted securities are all $\mathbb{Q}$-martingales. This is just a more precise name for what we call the martingale measure.

Already we have examples of the binomial trees and the continuous-time Black–Scholes model. Both of these are complete markets with an EMM. We have not found an arbitrage opportunity, but neither are we sure that one might not exist.

In both the binomial tree and Black–Scholes models we found there was one and only one EMM, and we were able to hedge claims. Even more so in the multiple stock models (section 6.3). There we could find a market price of risk γ_t but it (and so $\mathbb{Q}$ too) was only unique if the volatility matrix Σ_t was invertible. And it was exactly that invertibility which lets us hedge.

> **Arbitrage-free and completeness theorem (Harrison and Pliska)**
> Suppose we have a market of securities and a numeraire bond. Then
>
> (1) the market is arbitrage-free if and only if there is at least one EMM $\mathbb{Q}$; and
>
> (2) in which case, the market is complete if and only if there is exactly one such EMM $\mathbb{Q}$ and no other.

197

This simple yet powerful theorem makes sense of our experience.

In the HJM bond-market model, these conditions were also visible. The model demands that the forward rate drift $\alpha(t,T)$ satisfied

$$\alpha(t,T) = \sum_{i=1}^{n} \sigma_i(t,T) \left(\gamma_i(t) - \Sigma_i(t,T)\right),$$

for some previsible processes $\gamma_i(t)$. This ensures that there is an EMM $\mathbb{Q}$, and γ is the market price of risk. We now see that this is to make sure that the model is arbitrage-free.

The other key HJM condition is that the volatility matrix

$$\left(\Sigma_i(t,T_j)\right)_{i,j=1}^{n}$$

is non-singular for all sequences of dates $T_1 < \ldots < T_n$, and for all t less than T_1, which means there is only one viable price of risk in the market. This is sufficient (but actually slightly more than necessary) for the EMM to be unique, and consequently for the market to be complete.

It is worth getting a feel of why this theorem works. Although the technical details and exact definitions are passed over, the structure of the following can be proved rigorously.

Martingales mean no arbitrage

A martingale is really the essence of a lack of arbitrage. The governing rule for a $\mathbb{Q}$-martingale M_t is that

$$\mathbb{E}_{\mathbb{Q}}(M_t | \mathcal{F}_s) = M_s.$$

In other words, its future expectation, given the history up to time s, is just its current value at time s. The martingale is not 'expected' to be either higher or lower than its present value. An arbitrage opportunity, on the other hand, is a one-way bet which is certain to end up higher than it started.

Suppose we have a potential arbitrage opportunity contained in the self-financing portfolio strategy (ϕ, ψ). (Assuming for simplicity a two security market of stock S_t and bond B_t.) Then its value at time t is

$$V_t = \phi_t S_t + \psi_t B_t,$$

and it satisfies the self-financing equation

$$dV_t = \phi_t \, dS_t + \psi_t \, dB_t.$$

We can calculate the discounted value of the portfolio $E_t = B_t^{-1} V_t$, and then

$$dE_t = \phi_t \, dZ_t,$$

where Z_t is the discounted stock price $B_t^{-1} S_t$ which is a $\mathbb{Q}$-martingale.

Suppose now that the strategy does start with zero value ($V_0 = 0$) and finishes with a non–negative payoff ($V_T \geqslant 0$). Can this really be an arbitrage opportunity? Crucially, E_t is a $\mathbb{Q}$-martingale because Z_t is. And so

$$\mathbb{E}_{\mathbb{Q}}(E_T) = \mathbb{E}_{\mathbb{Q}}(E_T | \mathcal{F}_0) = E_0 = V_0 = 0.$$

But $V_T \geqslant 0$ and (because $B_T^{-1} > 0$) so is $E_T \geqslant 0$. But the $\mathbb{Q}$-expectation of E_T is zero, so the only possible value that E_T can take is zero too.

From which is it clear that V_T is zero as well. Any strategy can make no more than nothing from nothing. A martingale is essentially a 'fair game' and any strategy which involves only playing fair games cannot guarantee a profit.

Or in our language, if an EMM exists, there are no arbitrage opportunities.

Hedging means unique prices

If we can hedge, then there can only be at most one EMM.

To see this, suppose that we could hedge, but that there are two different EMMs $\mathbb{Q}$ and $\mathbb{Q}'$.

For any event A in the history $\mathcal{F}_T$, the digital–like claim which pays off the cash bond value at time T if A has happened has payoff $X = B_T I_A$. (The *indicator function* I_A takes the value 1 if the event A happens, and zero otherwise.) This is a valid derivative, so it must be hedgeable. (We assumed that we could hedge all claims.) So there must be a self-financing portfolio (ϕ, ψ) which hedges X, with value

$$V_t = \phi_t S_t + \psi_t B_t.$$

As usual the discounted claim $E_t = B_t^{-1} V_t$ satisfies

$$dE_t = \phi_t \, dZ_t,$$

where Z_t is the discounted stock price $B_t^{-1} S_t$. Now Z_t is both a $\mathbb{Q}$-martingale and a $\mathbb{Q}'$-martingale as both $\mathbb{Q}$ and $\mathbb{Q}'$ are EMMs. So also must E_t be. And from that, we see

$$E_0 = \mathbb{E}_{\mathbb{Q}}(E_T) = \mathbb{E}_{\mathbb{Q}'}(E_T).$$

But E_T is just the indicator function of the event A, I_A, and so $E_0 = \mathbb{Q}(A) = \mathbb{Q}'(A)$. The two measures $\mathbb{Q}$ and $\mathbb{Q}'$ which were trying to be different actually give the same likelihood for the event A. As A was completely general, the two measures agree completely, and thus $\mathbb{Q} = \mathbb{Q}'$. If any two EMMs are identical, then there can only really be one EMM.

Harrison and Pliska

We have only proved each result in one direction. We showed that if there was an EMM there was no arbitrage, but did not show that if there is no arbitrage then there actually is an EMM. Also we proved that hedging can only happen with a unique EMM, but not that the uniqueness of the EMM forced hedging to be possible.

The full and rigorous proofs of all these results in the discrete-time case are in the paper 'Martingales and stochastic integrals in the theory of continuous trading' by Michael Harrison and Stanley Pliska, in *Stochastic Processes and their Applications* (see appendix 1 for more details). For the continuous case and more advanced models, there has been other work, notably by Delbaen and Schachermayer. But the increasing technicality of this should not stand in the way of an appreciation of the remarkable insight of Harrison and Pliska.

Appendix 1
Further reading

The longer a list of books is, the fewer will actually be referred to. The lists below have been kept short, in the hope that in this case less choice is more.

Probability and stochastic calculus books

- *A first course in probability*, Sheldon Ross, Macmillan (4th edition 1994, 420 pages)
- *Probability and random processes*, Geoffrey Grimmett and David Stirzaker, Oxford University Press (2nd edition 1992, 540 pages)
- *Probability with martingales*, David Williams, Cambridge University Press (1991, 250 pages)
- *Continuous martingales and Brownian motion*, Daniel Revuz and Mark Yor, Springer (2nd edition 1994, 550 pages)
- *Diffusions, Markov processes, and martingales: vol. 2 Itô calculus*, Chris Rogers and David Williams, Wiley (1987, 475 pages)

These books are arranged in increasing degrees of technicality and depth (with the last two being at an equivalent level) and contain the probabilistic material used in chapters one, two and three. Ross is an introduction to the basic (static) probabilistic ideas of events, likelihood, distribution and expectation. Grimmett and Stirzaker contain that material in their first half, as well as the development of random processes including some basic material on martingales and Brownian motion.

Probability with martingales not only lays the groundwork for integration, (conditional) expectation and measures, but also is an excellent introduction to martingales themselves. There is also a chapter containing a simple representation theorem and a discrete-time version of Black–Scholes.

Both Revuz and Yor, and Rogers and Williams provide a detailed technical coverage of stochastic calculus. They both contain all our tools; stochastic differentials, Itô's formula, Cameron–Martin–Girsanov change of measure, and the representation theorem. Although dense with material, a reader with background knowledge will find them invaluable and definitive on questions of stochastic analysis.

Financial books

- *Options, futures, and other derivative securities*, John Hull, Prentice-Hall (2nd edition 1993, 490 pages)
- *Dynamic asset pricing theory*, Darrell Duffie, Princeton University Press (1992, 300 pages)
- *Option pricing: mathematical models and computation*, Paul Wilmott, Jeff Dewynne and Sam Howison, Oxford Financial Press (1993, 450 pages)

Hull is a popular book with practitioners, laying out the various real-world options contracts and markets before starting his analysis. A number of models are discussed, and numerical procedures for implementation are also included. The chapter-by-chapter bibliographies are another useful feature.

Duffie is a much more mathematically rigorous text, but still accessible. He contains sections on equilibrium pricing and optimal portfolio selection as well as a treatment of continuous-time arbitrage-free pricing along the same lines as this book. For readers with mathematical backgrounds, it is a good read.

Oxford Financial Press's volume comes at the subject purely from a differential equation framework without using stochastic techniques. Eventually, many pricing problems become differential equation problems, but unless a reader has experience in this area, it is not necessarily the best place to start from.

Chapter four: pricing market securities

Some notable journal papers include:
- The pricing of options and corporate liabilities, F Black and M Scholes, *Journal of Political Economy*, **81** (1973), 637–654.

- Theory of rational option pricing, R C Merton, *Bell Journal of Economics and Management Science*, **4** (1973), 141–183.
- Foreign currency option values, M B Garman and S W Kohlhagen, *Journal of International Money and Finance*, **2** (1983), 231–237.
- *Options markets*, J C Cox and M Rubinstein, Prentice-Hall (1985, 500 pages).
- Two into one, M Rubinstein, *RISK*, (May 1991), p. 49.

The Black–Scholes paper is now of historical interest, but it is still fascinating to see how the subject began, though the paper should be read for its insights, not the technical detail. At the time they were as concerned with pricing the stock of companies with outstanding liabilities (such as corporate bonds or warrants) as they were about options and derivatives.

Merton provides a more rigorous treatment, contemporaneously with Black–Scholes, and makes extension to dividend-paying stocks and a barrier option. Garman and Kohlhagen described foreign exchange options, whilst Cox and Rubinstein contain some exotic option formulas, amongst much else. The Rubinstein paper from *RISK* is concerned with quantos and cross-currency options.

Chapter five: interest rates

In the interest-rate setting, Heath–Jarrow–Morton is as seminal as Black–Scholes. By focusing on forward rates and especially by giving a careful stochastic treatment, they produced the most general (finite) Brownian interest-rate model possible. Other models may claim differently, but they are just HJM with different notation. The paper itself repays reading and re-reading.

- Bond pricing and the term structure of interest rates: a new methodology for contingent claims valuation, David Heath, Robert Jarrow and Andrew Morton, *Econometrica*, **60** (1992), 77–105.

In addition to the HJM paper, notable papers on the various interest-rate market models include

- Term structure movements and pricing interest rate contingent claims, T S Y Ho and S-B Lee, *Journal of Finance*, **41** (1986), 1011–1029.
- An equilibrium characterization of the term structure, O A Vasicek, *Journal of Finance*, **5** (1977), 177–188.

- Pricing interest rate derivative securities, J Hull and A White, *The Review of Financial Studies*, **3** (1990), 573–592.
- A theory of the term structure of interest rates, J C Cox, J E Ingersoll and S A Ross, *Econometrica*, **53** (1985), 385–407.
- Bond and option pricing when short rates are lognormal, F Black and P Karasinski, *Financial Analysts Journal*, (July–August 1991), 52–59.
- The market model of interest rate dynamics, A Brace, D Gatarek and M Musiela, *UNSW Preprint*, Department of Statistics S95-2.
- Which model for the term-structure of interest rates should one use?, L C G Rogers, in *Mathematical Finance* (ed. M H A Davis, D Duffie, *et al.*), IMA Volume 65, Springer-Verlag, 93–116.

The last of these is a review paper of models and their properties, whilst the others describe separately all the major models considered in the chapter.

Chapter six: bigger models

- Martingales and stochastic integrals in the theory of continuous trading, Michael Harrison and Stanley Pliska, *Stochastic Processes and their Applications*, **11** (1981), 215–260.
- The fundamental theorem of asset pricing, F Delbaen and W Schachermayer, *Mathematische Annalen*, **300** (1994), 463–520.
- The valuation of options for alternative stochastic processes, J C Cox and S A Ross, *Journal of Financial Economics*, **3** (1976), 145–166.

Harrison and Pliska made the next step forward by linking, in a general framework, the absence of arbitrage to the existence of a martingale measure, and showing that the ability to hedge depended on there only being one such measure. That this idea still underpins much of financial mathematics today is a demonstration of the importance of the paper.

Delbaen and Schachermayer go over similar ground but in a much more technical way to deal with the particular problems of continuous-time processes, including discontinuous processes. Cox and Ross cover option pricing for models more general than Black–Scholes, including those paying dividends.

Appendix 2
Notation

Notation can be divided naturally into three parts: lower case (generally deterministic), upper case (generally random), and Greek.

Lower case

a	a (real) parameter
c	a constant; coupon rate
$\frac{d\mathbb{Q}}{d\mathbb{P}}$	Radon–Nikodym derivative of $\mathbb{Q}$ with respect to $\mathbb{P}$
dt	infinitesimal time increment
dW_t	infinitesimal Brownian increment
f	a function
$f_{\mathbb{P}}(x)$	probability density function of the law $\mathbb{P}$
$f(t,T)$	bond forward rates
g	a function
$g(x,t,T)$	the function $\left(-\log P(t,T) \mid r_t = x\right)$
i	an integer
j	an integer
k	contract strike/exercise price; an integer; an offset
n	an integer
$n[t]$	number of dividend payments made by time t
p, p_j	a probability
q, q_j	a probability
r	constant interest rate
r_t	variable interest rate process; instantaneous rate

s	initial stock price, alternative time variable
s_j	possible value for the discrete stock process
t	time
u	foreign currency interest rate; real variable
x	a real variable; horizontal axis variable
$x_i(t)$	time-dependent factor of volatility surface
$y_i(T)$	maturity-dependent factor of volatility surface

Upper case

A	an event; a constant
A_t	HJM volatility matrix
B_i, B_t	bond price process
$B(t, T)$	solution of a Riccati equation
C_t	foreign exchange rate; coupon bond price; numeraire
D_i	financing gap
D_t	foreign currency cash bond
$D(t, T)$	solution of a Riccati equation
$\mathbb{E}$	expectation operator
$\mathbb{E}_{\mathbb{P}}$	expectation under the measure $\mathbb{P}$
E_t	discounted portfolio value process
F	forward price
$F_s(t, T)$	forward price at time s for $P(t, T)$
F_Q	quanto forward price
$\mathcal{F}_i$	history of discrete stock-price process up to tick-time i
$\mathcal{F}_t$	history of Brownian motion up to time t
I_A	indicator function of the event A
$I(t)$	sequence number of next coupon payment
K	option strike price
$L(T)$	LIBOR rate
$L(t, T)$	forward LIBOR rate
M_t	a martingale
N_t	a martingale
$\mathbb{N}$	the set of non-negative integers $\{0, 1, 2, \ldots\}$
$N(\mu, \sigma^2)$	a normal random variable with mean μ and variance σ^2
P	hypothetical discrete derivative price
$\mathbb{P}$	a probability measure
$\mathbb{P}_T$	forward measure

Notation

$P(t,T)$	bond prices
$\mathbb{Q}$	a probability measure
$\mathbb{R}^n$	the n-dimensional real vector space
$R(t,T)$	bond yield surface
S_i, S_t	stock price process
$\tilde{S}_t$	tradable asset price
$S_t^1 \ldots S_t^n$	stock price processes
T	maturity/exercise time of a derivative
T_i	coupon payment times
U_t	foreign currency derivative value process
V	derivative value
V_t	derivative value process
$V(s,T)$	Black–Scholes option price
$W_n(t)$	random walk
W_t	Brownian motion
$\tilde{W}_t$	Brownian motion
$W_t^1, \ldots, W_t^n$	independent Brownian motions
X	random variable; claim value of a derivative
X_i	sequence of random variables
X_t	a stochastic process
Y_t	a stochastic process
$Y_i(t,T)$	integral of y_i over $[t,T]$
Z	a (normal) random variable
Z_i, Z_t	discounted stock-price process
$Z(t,T)$	discounted bond prices
$\tilde{Z}_t$	discounted tradable asset price

Greek case

α	a real parameter
$\alpha(t,T)$	forward rate drift
$\beta(t,T)$	a function of two variables (Vasicek model)
γ_t	change of measure drift; market price of risk
$\gamma_i(t,T)$	BGM volatility surface
δ	dividend yield; coupon payment interval
δt	a small time increment
$\delta s_i, \ \delta n_i$	branch widths

ΔS_i, ΔV_i	change in value across δt of S_i, V_i, etc
ζ_t	change of measure process
θ	a real variable
θ_t	deterministic drift function
λ	a real parameter
μ	constant stock drift
μ_t	variable stock drift process
ν_t	stock drift process
π_i	path probability
Π_i	portfolio
ρ	correlation
$\bar{\rho}$	the orthogonal complement $\sqrt{1 - \rho^2}$
ρ_t	volatility process
σ	constant stock volatility
σ_1, σ_2	stock volatilities
σ_t	variable stock volatility process
$\sigma(t, T)$	forward rate volatility surface
$\sigma_i(t, T)$	multi-factor forward rate volatility surface
$\bar{\sigma}$	term volatility
Σ_t	volatility matrix
$\Sigma(t, T)$	bond price volatilities
τ	time horizon; maturity date; stopping time
$\phi_t, \tilde{\phi}_t$	stock-holding strategy; representation theorem integrand
Φ	normal distribution function: $\Phi(x) = \mathbb{P}(N(0, 1) \leqslant x)$
ψ_t	bond-holding trading strategy
ω	a sample path

Appendix 3
Answers to exercises

2.1 The value (at maturity) of the futures contract is the stock price less the exercise price. If the process moved to node 2, the stock price is s_2, so the future's value is $f(2) = s_2 - k$. The other node is similar. Then, recalling that q is $(s_1 \exp(r\, \delta t) - s_2)/(s_3 - s_2)$,

$$V = \exp(-r\, \delta t)\big((1-q)(s_2 - k) + q(s_3 - k)\big)$$
$$= \exp(-r\, \delta t)\left(s_2 \frac{s_3 - s_1 e^{r\, \delta t}}{s_3 - s_2} + s_3 \frac{s_1 e^{r\, \delta t} - s_2}{s_3 - s_2} - k\right).$$

This is equal to $e^{-r\, \delta t}(s_1 e^{r\, \delta t} - k)$, which can be simplified to give $V = s_1 - k e^{-r\, \delta t}$. The only strike price which gives zero present value to the future is $k = s_1 \exp(r\, \delta t)$.

Table A.1 Option and portfolio development – in the money

Time i	Last Jump	Stock Price S_i	Option Value V_i	Stock Holding ϕ_i	Bond Holding ψ_i
0	–	100	50	–	–
1	up	120	75	1.25	−75
2	up	140	100	1.25	−75
3	down	100	100	0.00	100

2.2 The progression, in the first scenario, of the stock price and hedging strategy is laid out in table A.1, and the option claim tree is shown in figure A.1. The corresponding table in the other case is table A.2. The price of the option at time 0 is 50. We can represent the hedging strategy ϕ with a tree up to time 2, each of whose nodes give the amount of stock which should be held from that time for the next period. The tree is given in figure A.2.

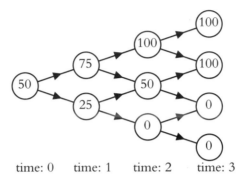

Figure A.1 The option claim tree for a digital payoff

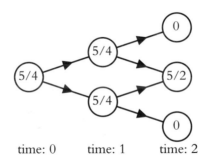

Figure A.2 The hedging strategy tree

Table A.2 Option and portfolio development – out of the money

Time i	Last Jump	Stock Price S_i	Option Value V_i	Stock Holding ϕ_i	Bond Holding ψ_i
0	–	100	50	–	–
1	down	80	25	1.25	−75
2	up	100	50	1.25	−75
3	down	80	0	2.50	−200

2.3 Table A.3 has the calculations which show that $\mathbb{E}_{\mathbb{Q}}(S_2|\mathcal{F}_i)$ is identical to S_i. That S_i is a $\mathbb{Q}$-martingale is now immediate from the remarks following the tower law.

Table A.3 Conditional expectation against filtration value

Expectation	Filtration	Value
$\mathbb{E}_{\mathbb{Q}}(S_2|\mathcal{F}_0)$	$\{1\}$	$\frac{1}{3} \cdot \frac{2}{5} \cdot 180 + \frac{1}{3} \cdot \frac{3}{5} \cdot 80 + \frac{2}{3} \cdot \frac{2}{3} \cdot 72 \ldots$ $+ \frac{2}{3} \cdot \frac{1}{3} \cdot 36 = 80$
$\mathbb{E}_{\mathbb{Q}}(S_2|\mathcal{F}_1)$	$\{1,3\}$ $\{1,2\}$	$\frac{2}{5} \cdot 180 + \frac{3}{5} \cdot 80 = 120$ $\frac{2}{3} \cdot 72 + \frac{1}{3} \cdot 36 = 60$
$\mathbb{E}_{\mathbb{Q}}(S_2|\mathcal{F}_2)$	$\{1,3,7\}$ $\{1,3,6\}$ $\{1,2,5\}$ $\{1,2,4\}$	180 80 72 36

3.1 No. The increments are wrong (or equivalently the conditional distributions are wrong). The increment $X_{s+t} - X_s$ is a normal with variance $t - 2s(\sqrt{1+t/s} - 1)$, which is not t, and the increment is not independent of X_s.

3.2 Yes. The increment $X_{s+t} - X_s$ is the sum of a normal $N(0, t\rho^2)$ and an independent normal $N(0, t(1-\rho^2))$, which is equal to a normal $N(0,t)$ in total. The increment is certainly independent of both histories $(W_u : u \leqslant s)$ and $(\tilde{W}_u : u \leqslant s)$, and hence also independent of the history $(X_u : u \leqslant s)$.

3.3 As S_T is distributed as a normal random variable with mean μT and variance $\sigma^2 T$, the probability that this is negative is the same as the probability that a normal $N(0,1)$ random variable is less than $-\mu\sqrt{T}/\sigma$. This probability is positive.

3.4 $dX_t = \exp(W_t) \, dW_t + \frac{1}{2}\exp(W_t) \, dt = X_t \, dW_t + \frac{1}{2}X_t \, dt$.

3.5 $X_t = X_0 \exp(\sigma W_t + \int_0^t \mu_s \, ds - \frac{1}{2}\sigma^2 t)$.

3.6 We know that the processes can be written as $dB_t = \beta_t \, dt$ and $dX_t =$

$\sigma_t\, dW_t + \mu_t\, dt$. Then

$$d(B_t X_t) = \tfrac{1}{2} d((B_t + X_t)^2 - B_t^2 - X_t^2),$$

which by Itô is equal to

$$(B_t + X_t)(dB_t + dX_t) + \tfrac{1}{2}\sigma_t^2\, dt - B_t\, dB_t - X_t\, dX_t - \tfrac{1}{2}\sigma_t^2\, dt.$$

This itself simplifies to the desired answer. Alternatively, an application of the product rules gives an immediate solution.

3.7 The formula will hold (by definition) for $t = 2$. At time $t = 1$, if the first jump was 'up', then $\frac{dQ}{dP}$ will be $q_1 q_2 / p_1 p_2$ with $\mathbb{P}$-probability p_2, and $q_1 \bar{q}_2 / p_1 \bar{p}_2$ with $\mathbb{P}$-probability $\bar{p}_2$. The (conditional) expectation is then $q_1 q_2 / p_1 + q_1 \bar{q}_2 / p_1 = q_1 / p_1$. This is exactly what ζ_1 is, if the first jump were up. The case where the first jump is down is similar. At time $t = 0$, $\mathbb{E}_{\mathbb{P}}(\frac{dQ}{dP} | \mathcal{F}_0) = \mathbb{E}_{\mathbb{P}}(\frac{dQ}{dP}) = \mathbb{E}_{\mathbb{Q}}(1) = 1$. This matches ζ_0, as we desired.

3.8 We have to prove the result for all s and t, such that $s \leqslant t$, and for all possible X_t. We note firstly that the $s = 0$ case follows immediately from the fact that $\mathbb{E}_{\mathbb{Q}}(X) = \mathbb{E}_{\mathbb{P}}(\frac{dQ}{dP} X)$, and that ζ_t is $\frac{dQ}{dP}$ at time t. The $s = t$ case is also trivial, as both sides of the result are simply X_t. So in fact, we need only check the case $s = 1$, $t = 2$, for four cases of X_2. Take the case where X_2 is the digital claim which only pays off 1 if the process goes up twice. If the first jump was down, then both sides of the result are zero, because X_2 and $\zeta_2 X_2$ must both be zero. If the first jump was up, then the left-hand side is

$$\mathbb{E}_{\mathbb{Q}}(X_2 | \mathcal{F}_1) = q_2(1) + \bar{q}_2(0) = q_2.$$

The right-hand side is

$$\zeta_1^{-1} \mathbb{E}_{\mathbb{P}}(\zeta_2 X_2 | \mathcal{F}_1) = \frac{p_1}{q_1}\left(p_2\left(\frac{q_1 q_2}{p_1 p_2} 1 \right) + \bar{p}_2(0) \right) = q_2.$$

Similarly, we can check the digital claims where X_2 pays off only after an up-down, a down-up, and a down-down. This completes the verification.

3.9 The equivalence of (ii) and (ii)' is immediate from the boxed theorem about identifying normals. So too is the equivalence of (iii) and (iii)', as long as we note that the right-hand side of (iii)' is independent of the history $\mathcal{F}_s$.

We can also note that (ii)' is just a special case of (iii)' when $s = 0$. So it is enough just to prove (iii)'. The left-hand side is

$$\mathbb{E}_{\mathbb{Q}}\left(\exp\left(\theta(W_{t+s} - W_s + \gamma t)\right) \mid \mathcal{F}_s\right)$$
$$= \zeta_s^{-1}\mathbb{E}_{\mathbb{P}}\left(\zeta_{t+s} \exp\left(\theta(W_{t+s} - W_s + \gamma t)\right) \mid \mathcal{F}_s\right).$$

Here, $\zeta_t = \mathbb{E}_{\mathbb{P}}(\frac{d\mathbb{Q}}{d\mathbb{P}} \mid \mathcal{F}_t)$. By property (iii) of the $\mathbb{P}$-Brownian motion W, $\frac{d\mathbb{Q}}{d\mathbb{P}} = \exp(-\gamma W_t - \frac{1}{2}\gamma^2 T)\exp(-\gamma(W_T - W_t))$, where $W_T - W_t$ is a normal $N(0, T-t)$ independent of $\mathcal{F}_t$, from which it follows that $\zeta_t = \exp(-\gamma W_t - \frac{1}{2}\gamma^2 T)\exp(\frac{1}{2}\gamma^2(T-t))$, which is just

$$\zeta_t = \exp(-\gamma W_t - \tfrac{1}{2}\gamma^2 t).$$

Thus the left-hand side of (iii)' becomes

$$\exp(\theta\gamma t - \tfrac{1}{2}\gamma^2 t)\mathbb{E}_{\mathbb{P}}\left(\exp\left((\theta - \gamma)(W_{t+s} - W_s)\right) \mid \mathcal{F}_s\right).$$

Again using the fact that $W_{t+s} - W_s$ is a normal $N(0, t)$ independent of $\mathcal{F}_s$, the expectation part of the above is $\exp(\frac{1}{2}(\theta - \gamma)^2 t)$, which means that the whole expression actually is $\exp(\frac{1}{2}\theta^2 t)$, as desired.

3.10 If $\gamma = 0$, then $X_t = W_t$, which we know is a martingale by example (2). If, however, γ is not zero, then

$$\mathbb{E}(X_t|\mathcal{F}_s) = \mathbb{E}(W_t|\mathcal{F}_s) + \gamma t = W_s + \gamma t = X_s + \gamma(t - s).$$

If γ is not zero, then X is not a martingale, because of the extra term $\gamma(t - s)$ above.

3.11 The function σ is bounded, in that there is a constant K such that $|\sigma(t, w)| \leq K$, for all $t \leq T$, and for all w in Ω. Then $\exp(\frac{1}{2}\int_0^T \sigma_s^2 \, ds)$ is bounded above by $\exp(\frac{1}{2}K^2 T)$ for all w, so its expectation is also bounded by that constant. By the second collector's guide box, the local martingale X must also be a martingale.

Appendices

3.12 To differentiate $V_t = W_t^2 - t$, we can treat the two terms W_t^2 and $-t$ separately. The first of these can be handled by applying Itô's formula to the case $f(x) = x^2$, with $X_t = W_t$. Then

$$d\big(f(W_t)\big) = f'(W_t)\,dW_t + \tfrac{1}{2}f''(W_t)\,dt = 2W_t\,dW_t + dt.$$

Also $d(-t) = -dt$, so we can deduce that $dV_t = 2W_t\,dW_t$. In fact, V_t is a proper martingale, because if we let X be $(\int_0^T W_t^2\,dt)^{\frac{1}{2}}$, then it is enough to show (collector's guide) that $\mathbb{E}(X) < \infty$. In fact

$$\big(\mathbb{E}(X)\big)^2 \leqslant \mathbb{E}(X^2) = \int_0^T \mathbb{E}(W_t^2)\,dt = \tfrac{1}{2}T^2.$$

3.13 If we let L_t be the logarithm of Z_t, then $L_t = \sigma W_t + (\mu - r)t$, and so $dL_t = \sigma\,dW_t + (\mu - r)\,dt$. The expression then follows from Itô.

3.14 If we change variables in the integral to $v = -(x + \tfrac{1}{2}\sigma^2 T)/\sigma\sqrt{T}$, then V_0 is

$$V_0 = \frac{1}{\sqrt{2\pi}} \int_{-\infty}^a \left(s e^{-\sigma\sqrt{T}v - \frac{1}{2}\sigma^2 T} - k e^{-rT} \right) e^{-\frac{1}{2}v^2}\,dv,$$

where a is the constant $(\log \tfrac{s}{k} + (r - \tfrac{1}{2}\sigma^2)T)/\sigma\sqrt{T}$. By writing $\exp(-\sigma\sqrt{T}v - \tfrac{1}{2}\sigma^2 T - \tfrac{1}{2}v^2)$ as $\exp\big(-\tfrac{1}{2}(v + \sigma\sqrt{T})^2\big)$, we can rewrite the integral as

$$V_0 = \frac{s}{\sqrt{2\pi}} \int_{-\infty}^{a+\sigma\sqrt{T}} e^{-\frac{1}{2}v^2}\,dv - \frac{ke^{-rT}}{\sqrt{2\pi}} \int_{-\infty}^a e^{-\frac{1}{2}v^2}\,dv.$$

This can then be evaluated as

$$V_0 = s\Phi(a + \sigma\sqrt{T}) - ke^{-rT}\Phi(a),$$

which is the expression we sought.

3.15 We needed to know that the drift was constant. So far, we can only cancel constant drifts with our three-step plan. Later (section 6.1) we will generalise this, but for the moment we need the drift to be constant, even if we are indifferent to which constant it is.

3.16 Simply evaluate the Black–Scholes formula with $s = \$10$, $k = \$12$, $\mu = 0.15$, $\sigma = 0.20$, $r = 0.05$ and $T = 1$. The option value is $\$0.325$.

3.17 In this case X is \$1 if $S_T > \$10$, and is zero otherwise, where T is 1. Hence by the derivative pricing formula

$$V_0 = \mathbb{E}_{\mathbb{Q}}(B_T^{-1}X) = e^{-rT}\mathbb{Q}(S_T > \$10) = e^{-rT}\Phi\left(\frac{rT - \frac{1}{2}\sigma^2 T}{\sigma\sqrt{T}}\right).$$

This has the numerical value of \$0.532.

4.1 (i) Discounted, the asset is $Z_t = B_t^{-1}X_t = \exp(2\sigma\tilde{W}_t + (r - \sigma^2)t)$. Its SDE is $dZ_t = Z_t(2\sigma\,d\tilde{W}_t + (r + \sigma^2)\,dt)$, which has a non-vanishing drift term. So Z_t is not a $\mathbb{Q}$-martingale, and thus X_t is not a tradable asset. (ii) In this case, the discounted asset is $Z_t = B_t^{-1}X_t = \exp(-\alpha\sigma\tilde{W}_t - \alpha r t)$. Given that, $\alpha r = \frac{1}{2}(\alpha\sigma)^2$, the SDE of Z_t is $dZ_t = Z_t(-\alpha\sigma\,d\tilde{W}_t)$, which is a $\mathbb{Q}$-martingale. So X_t is tradable.

4.2 Replace each $dW_i(t)$ by $d\tilde{W}_i(t) - \gamma_i(t)$ and substitute into the SDEs for dY_t and dZ_t and see that the drift terms vanish.

4.3 The only difference between this example and the sterling case in the text is that the exchange rate is the other way round. Before we had the sterling/dollar rate (the worth of the local currency in domestic terms), and here we have the dollar/yen rate (the worth of the domestic currency in local terms). We should really be working with C_t^{-1} instead of C_t, but the only difference is that the sign of the correlation changes. Thus the forward price is $F_0 = \exp(\rho\sigma_1\sigma_2)F$, and not $\exp(-\rho\sigma_1\sigma_2)F$, where F is the local currency forward $F = e^{uT}S_0$. As exchange rates tend to quote the 'big' number, the sign of ρ needed in any particular instance depends on the actual pair of currencies in question.

Appendix 4
Glossary of technical terms

Adapted a process which depends only on the current position and past movements of the driving processes. It is unable to see into the future

American call option a call option which can be exercised at any time up to the option expiry date

Arbitrage the making of a guaranteed risk-free profit with a trade or series of trades in the market

Arbitrage free a market which has no opportunities for risk-free profit

Arbitrage price the only price for a security that allows no arbitrage opportunity

Autoregressive of a positive process, that it is *mean-reverting*

Average the arithmetic mean of a sample

Bank account process an account which is continuously compounded at the prevailing instantaneous rate, and behaves like the *cash bond*

Binomial process a process on a binomial tree

Binomial representation theorem a discrete-time version of the *martingale representation theorem* on the binomial tree

Binomial tree a tree, each of whose nodes branches into two at the next stage

Black–Scholes a stock market model with an analytic option pricing formula

Bonds
interest bearing securities which can either make regular interest payments and/or a lump sum payment at maturity

Bond options
an option to buy or sell a bond at a future date

Brownian motion
the basic stochastic process formed by taking the limit of finer and finer random walks. It is a martingale, with zero drift and unit volatility, and is not Newtonian differentiable

Calculus
generally a formal system of calculation, in particular concerned with analysing behaviour in terms of infinitesimal changes of the variables. *Newtonian calculus* handles smooth functions, but not Brownian motion which requires the techniques of *stochastic calculus*. [From *calculus* (Lat.), a pebble used in an abacus]

Call option
the option to buy a security at/by a future date for a price specified now

Cameron–Martin–Girsanov theorem
a result which interprets equivalent change of measure as changing the drift of a Brownian motion

Cap
a contract which periodically pays the difference between current interest rate returns and a rate specified at the start, only if this difference is positive. A cap can be used to protect a borrower against floating interest rates being too high

Caplet
an individual cap payment at some instant

Cash bond
a liquid continuously compounded bond which appreciates at the instantaneous interest rate

Central limit theorem
a statistical result, which says that the average of a sample of IID random variables is asymptotically normally distributed

Change of measure
viewing the same stochastic process under a different set of likelihoods, changing the probabilities of various events occurring

Claim
a payment which will be made in the future according to a contract

Commodity
a real thing, such as gold, oil or frozen concentrated orange juice

Complete market
a market in which every claim is hedgable

Conditional distribution	the distribution of a random variable conditional on some information $\mathcal{F}$, such as $\mathbb{P}(X \leqslant x \vert \mathcal{F})$
Conditional expectation	taking an expectation given some history as known. For instance the conditional expectation of the number of heads obtained in three tosses, given that the first toss was heads, is two; whereas the unconditioned expectation is only one and a half. Written $\mathbb{E}(\cdot \vert \mathcal{F}_t)$, for conditioning on the history of the process up to time t
Contingent claim	a claim whose amount is determined by the behaviour of market securities up until the time it is paid
Continuous	a process or function which only changes by a small amount when its variable or parameter is altered infinitesimally
Continuous-time	a process which depends on a real-valued time parameter, allowing infinite divisibility of time
Continuously compounded	interest is compounded instantly, rather than annually or monthly, leading to exponential growth
Contract	an agreement under law between two principals, or counterparties
Correlation	a measure of the linear dependence of two random variables. If one variable gets larger as the other does, the correlation is positive, and negative if one gets larger as the other gets smaller. The limits of one and minus one correspond to exact dependence, whereas independent variables have zero correlation. Formally correlation is the *covariance* of the random variables divided by the square root of the product of their individual variances
Coupon	a periodic payment made by a bond
Covariance	a measure of the relationship of two random variables, the covariance is zero if the variables are independent (and *vice versa* in the case of jointly normal random variables). Formally the covariance of two variables is the expectation of their product less the product of their expectations
Cumulative normal integral	see *normal distribution function*

Currency	the monetary unit of a country or group of countries
Default free	there being no chance that the bond issuer will be unable to meet his financial undertakings (used theoretically)
Density	the probability density function f is the derivative (if it exists) of the distribution function of a continuous random variable. Intuitively, $f(x)\,dx$ is the probability that X lies in the interval $[x, x + dx]$. The function f is non-negative, integrates to one, and can be used to calculate expectations, and so forth, as

$$\mathbb{E}(X^2) = \int_{-\infty}^{\infty} x^2 f(x)\,dx$$

Derivative	a security whose value is dependent on (derived from) existing underlying market securities. See also *contingent claim*
Difference equation	the discrete analogue of a differential equation. For example, to find the sequence (x_n) which obeys

$$ax_{n+2} + bx_{n+1} + cx_n = d$$

Diffusion	a stochastic process which is the solution to a SDE
Digital	a derivative which pays off a fixed amount if a given future event happens, and nothing otherwise
Discount	scaling a future reward or cost down to reflect the importance of now over later
Discount bond	a bond which promises to make a lump sum payment at a future date, but until then is worth less than its face value
Discrete	taking distinct, separated values; such as from the sets $\mathbb{N}$ or $\{0, \delta t, 2\delta t, \ldots\}$
Distribution	of a random variable, the description of the likelihood of its every possible value
Distribution function	the (cumulative) distribution function F of a random variable is defined so that $F(x)$ is the probability that the random variable is no larger than x. The

Appendices

Distribution function (contd)   function F increases (weakly) from 0 to 1. If F is differentiable, then its derivative is the density

Dividends   regular but variable payments made by an equity

Doléans exponential   for a local martingale M_t, this is the solution of the SDE $dX_t = X_t\, dM_t$, which is another local martingale $X_t = \exp(M_t - \frac{1}{2}\int_0^t (dM_s)^2)$

Drift   the coefficient of the dt term of a stochastic process

Driftless   a process with constant zero drift

Equilibrium distribution   a distribution of a process which is stable under time evolution

Equities   stocks which make dividend payments

Equivalent martingale measure (EMM)   see *martingale measure*

Equivalent measures   two measures $\mathbb{P}$ and $\mathbb{Q}$ are equivalent if they agree on which events have zero probability

European call option   a call option which can be exercised or not only at the option exercise date. Compare with *American call option*

Exercise date   a set future date at which an option may be exercised or not

Exercise price   see *strike price*

Exotics   new derivative securities, which will quickly either become standard products or will sink without trace

Expectation   the mean of a random variable, which will be the limiting value of the average of an infinite number of identical trials. For a discrete and a continuous random variable (with density f) it is respectively

$$\mathbb{E}(X) = \sum_{n=0}^{\infty} n\, \mathbb{P}(X = n), \quad \mathbb{E}(X) = \int_{-\infty}^{\infty} x f(x)\, dx$$

Exponential Brownian motion   a process which is the exponential of a drifting Brownian motion

Exponential martingales   the *Doléans exponential* of a martingale, which itself is a (local) martingale

Filtration	the history, $(\mathcal{F}_t)_{t \geqslant 0}$, of a process, where $\mathcal{F}_t$ is the information about the path of the process up to time t
Fixed	of interest rates, that they are constant throughout the term of the contract
Floating	of interest rates, that they can move with the market over the term of the contract
Floor	a contract which periodically pays the difference between a rate specified at the start and current interest rate returns, only if this difference is positive. A floor can be used to protect a lender against floating interest rates being too low. See also *cap*
Floorlet	which is to floors as caplets are to caps
Foreign exchange	the market which prices one currency in terms of another
Forward	an agreement to buy or sell something at a future date for a set price, called the *forward price*
Forward rate	the forward price of instantaneous borrowing
Fractal	a geometrical shape which on a small-scale looks the same as the large-scale, only smaller. A straight line is a fractal of dimension one, and a Brownian motion path is a fractal of dimension 1.5
Future	a *forward* traded on an exchange
FX	abbreviation for *foreign exchange*
Gaussian process	a process, all of whose marginals are normally distributed, and all of whose joint distributions are jointly normal
Heath–Jarrow–Morton (HJM)	a model of the interest-rate market
Hedge	to protect a position against the risk of market movements
History	the information recording the path of a process
Identically distributed	of random variables, have the same probabilistic distribution
IID	abbreviation for Independent, Identically Distributed

Independent	of variables, none of which have any relation or influence on any of the others
Indicator function	a function of a set which is one when the argument lies in the set and zero when it is outside
Induction	a method of proof, involving the demonstration that the current case follows from the previous case, which itself then implies the next case, and so on
Instantaneous rate	the rate of interest paid on a very very short term loan
Instruments	tradable securities or contracts
Interest rate	the rate at which interest is paid
Interest rate market	the market which determines the *time value of money*
Itô's formula	a stochastic version of the 'chain rule' which expresses the volatility and drift of the function of a stochastic process in terms of the volatility and drift of the process itself and the derivatives of the function. If X_t has volatility σ_t and drift μ_t, then $Y_t = f(X_t)$ has volatility $f'(X_t)\sigma_t$ and drift $f'(X_t)\mu_t + \frac{1}{2}f''(t)\sigma_t^2$
Kolmogorov's strong law	see *strong law*
Law of the unconscious statistician	the result that if a random variable X has density f, then the expectation of $h(X)$ is $$\mathbb{E}\big(h(X)\big) = \int_{-\infty}^{\infty} h(x)f(x)\,dx$$
LIBOR	the London Inter-Bank Offer Rate. A daily set of interest rates for various currencies and maturities
Local martingale	a stochastic process which is driftless, but not necessarily a martingale
Log-drift	of a stochastic process X_t, the drift of $\log X_t$
Log-normal distribution	a random variable whose logarithm is normally distributed
Log-volatility	of X_t is the volatility of $\log X_t$, or equivalently the volatility of dX_t/X_t
Long	(of position) having a positive holding

Marginal the marginal distribution of a process X at time t is the distribution of X_t considered as a random variable in isolation. Two processes may be different, yet have exactly the same marginal distributions

Market a place for the exchanging of price information. Commonly situated in electronic space

Market maker (in UK) a dealer who is obligated to quote and trade at two-way prices

Market price of risk a standardised reward from risky investments in terms of extra growth rate

Markov of a process, meaning that its future behaviour is independent of its past, conditional on the present

Martingale a process whose expected future value, conditional on the past, is its current value. That is, $\mathbb{E}(M_t|\mathcal{F}_s)$ equals M_s for every s less than t

Martingale measure a measure under which a process is a martingale

Martingale representation theorem a result which allows one martingale to be written as the integral of a previsible process with respect to another martingale

Maturity the time at which a bond will repay its principal, or more generally the time at which any claim pays off

Mean synonym for *expectation*

Mean reversion the property of a process which ensures that it keeps returning to its long-term average

Measure a collection of probabilities on the set of all possible outcomes, describing how likely each one is

Multi-factor a market model which is driven by more than one Brownian motion

Newtonian calculus classical differential and integral calculus, relating to smooth or differentiable functions

Newtonian function a function which is smooth enough to have a classical (Newtonian) derivative

Node a point on a tree where branches start and finish

Noise a loose term for *volatility*

Normal distribution	a continuous distribution, parameterised by a mean μ and variance σ^2, written $N(\mu, \sigma^2)$ with density

$$f(x) = \frac{1}{\sqrt{2\pi\sigma^2}} \exp\left(-\frac{(x-\mu)^2}{2\sigma^2}\right)$$

Normal distribution function	the distribution function of the normal random variable, written $\Phi(x) = \mathbb{P}\big(N(0,1) \leqslant x\big)$
Numeraire	a basic security relative to which the value of other securities can be judged. Often the *cash bond*
ODE	abbreviation for Ordinary Differential Equation
Option	a contract which gives the right but not the obligation to do something at a future date
Ornstein–Uhlenbeck (O-U) process	a mean reverting stochastic process with SDE

$$dX_t = \sigma\, dW_t + (\theta - \alpha X_t)\, dt$$

Over-the-counter	an agreement concluded directly between two parties, without the mediation of an exchange
Path probability	the probability of a tree process taking a particular path through the tree. The probability will be the product of the probabilities of the individual branches taken
Payoff	a payment
PDE	abbreviation for Partial Differential Equation
Poisson process	a type of random process with discontinuities
Portfolio	a collection of security holdings
Position	the amount of a security held, which can either be positive (a long position) or negative (a short position)
Previsible	a stochastic process which is adapted and is either continuous or left-continuous with right-limits or is a limit of such processes
Principal	the face value that a bond will pay back at maturity
Probability	the chance of an event occurring

Process — a sequence of random variables, parameterised by time

Product rule — a result giving the stochastic differential of the product of two stochastic processes

Put-call parity — the observation that the worth of a call less the price of a put struck at the same price is the current worth of a forward

Quantos — cross-currency contracts, derivatives which pay off in another currency

Radon–Nikodym derivative — of one measure with respect to another is the relative likelihood of each sample path under one measure compared with the other

Random variable — a function of a sample space

Random walk — a discrete Markov process made up of the sum of a number of independent steps. A simple symmetric random walk is $\mathbb{N}$-valued and after each time step goes up one with probability $\frac{1}{2}$ and down one with probability $\frac{1}{2}$

Recombinant tree — a tree where branches can come together again

Replicating strategy — a *self-financing* portfolio trading strategy which hedges a claim precisely

Risk free — no chance of anything going wrong

Risk-neutral measure — a *martingale measure*

SDE — abbreviation for Stochastic Differential Equation

Security — a piece of paper representing a promise

Self-financing — a strategy which never needs to be topped up with extra cash nor can ever afford withdrawals

Semimartingale — a process which can be decomposed into a local martingale term and a drift term of finite variation

Share — (in UK) a stock or equity

Short — (of position) having a negative, or borrowed, holding

Short rate — see *instantaneous rate*

Single-factor — a market model which is driven by only one Brownian motion

Standard deviation   the square root of the variance

Stochastic   synonym for random

Stochastic calculus   a *calculus* for random processes, such as those involving Brownian motion terms

Stochastic process   a continuous process, which can be decomposed into a Brownian motion term and a drift term

Stock   a security representing partial ownership of a company

Stock market   a place for trading stocks

Strike price   the price at which an asset may be bought or sold under an option

Strong law   the result that the average of a sample of n IID random variables will converge to the mean of the distribution as n increases, given some technical conditions

Swaps   an agreement to make a series of fixed payments over time and receive a corresponding series of payments dependent on current interest rates, or *vice versa*

Swaption   an option to enter into a swap agreement at a future date

Taylor expansion   for Newtonian functions, the expression of the value of a function f near x in terms of the value of it and its derivatives at x, that is

$$f(x+h) = f(x) + hf'(x) + \tfrac{1}{2}h^2 f''(x) + \tfrac{1}{6}h^3 f'''(x) \ldots$$

Term structure   the relationship between the interest rates demanded on loans, and the length of the loans

Term variance   the variance of the logarithm of a security price over a time period, $\mathrm{Var}\big(\log(S_T/S_0)\big)$

Term volatility   the effective (annualised) volatility of an asset over a time period. Explicitly, its square is the *term variance* divided by the length of the term:
$\bar{\sigma}^2 = \mathrm{Var}\big(\log(S_T/S_0)\big)/T$

Time value of money   the difference between cash now, and cash later which is subject to a *discount*

Tower law	the result that $\mathbb{E}\big(\mathbb{E}(X|\mathcal{F}_t) \mid \mathcal{F}_s\big) = \mathbb{E}(X|\mathcal{F}_s)$, for $s < t$
Tradable	of an asset, that it can be traded either directly, or indirectly by trading a matching portfolio
Trading strategy	a continuous choice of portfolio, a choice which may depend on market movements
Transaction cost	a charge for buying or selling a security
Tree	a graph of nodes linked by branches which contains no closed loops or circuits
Underlying	a basic market security, such as stocks, bonds and currencies
Vanilla	of a product, the standard basic version
Variable coupon	periodic payments from a floating interest-rate contract
Variance	a measure of the uncertainty of a random variable. Formally, the expectation of its square less the square of its expectation, or equivalently the expected square of the difference between the random variable and its mean
Volatility	the amount of 'noise' or variability of a process, more precisely, the coefficient of the Brownian motion term of a stochastic process
Weak law	the result that the average of n IID random variables is increasingly less likely to be significantly different from the distribution mean as n increases
Wiener process	synonym for *Brownian motion*
With probability 1	of an event, having probability one of occurring. This is not quite the same as being guaranteed for sure, as, for example, a normal random variable can take the value zero, but with probability one it will not
Yield	the average interest rate offered by a bond
Yield curve	the graph of yield plotted against bond maturity
Zero coupon	a bond which does not make any payments until maturity

Index

Index